"Whilst the conceptualisation of edutainment is not new, see this unique edited text bringing together interconnected concepts such as planning, engagement, and the event experience giving readers a much deeper understanding into how edutainment can be applied in a multidisciplinary way to improve sustainability, create equity, and move EDI agendas forward, upon which the future of humanity depends."

– **Dr Allan Jepson**, *Hertfordshire Business School*

"Festivals and Edutainment is a vital addition to existing knowledge on Events Management theory and practice. It is your one stop shop for examining the theoretical and practical nuances of Education and Entertainment in the Festival typology of events. The blend of industry and research insights means the text lends itself to industry practitioners as well as taught units aligned to Events and Arts Management."

– **Dr Miriam Firth**, *School of Environment, Education and Development, Manchester Institute of Education*

"Decades, if not centuries later, we are still – only now – starting to understand the transformational power of events on the people that organize and experience them. We must do more to analyze how events move us, change us, and generate positive social change. By "we" I mean … everyone! Everyone involved in the conception, planning, delivery, and the long-tail legacies that events bequeath. That's why this book, focused on "edutainment", is special; a series of well-written and interesting cases that help to explicate these complex processes, applicable to all event stakeholders, from academics to practitioners."

– **Dr Mike Duignan**, *Rosen College of Hospitality Management, University of Central Florida*

"This pioneering volume provides a novel approach to the understudied theme of edutainment in events. It assembles a curated collection of contributions with a wide range of perspectives and incorporating different research methodologies. Important contemporary issues such as sustainability and equality are examined, and visitor and resident perspectives, edutainment experiences and emotions are highlighted. It therefore represents an important source book on edutainment in events for researchers and students."

– **Professor Greg Richards**, *Tilburg School of Social and Behavioural Sciences, Tilburg University*

"This book is the first to unpack the role of edutainment in festival experiences globally. While much festival research has focused on the operational elements of festival management, the concept of edutainment has received much less attention. Using a series of case studies of diverse festivals from around the world, the book examines how edutainment festival experiences are planned to engage audiences and how audiences experience edutain-ment. The case studies also explore how edutainment fosters co-creation leading to more inclusive and sustainable festivals. This text will be invalua-ble reading for advanced students of events and festivals as well as tourism and leisure students more broadly as well as festival practitioners."

— **Professor Kirsten Holmes**, *School of Management and Marketing,*
Curtin University

Festivals and Edutainment

As the first collection of studies to explore the use of edutainment within festival experiences, this book extends current knowledge and understanding of festival experiences. Relying on a series of international case studies, this book offers readers unique and important insights that emphasise the benefit of edutainment activities for enhanced audience learning, engagement, and festival satisfaction.

Although there is an ample amount of studies concerning festival experiences, as well as the use of edutainment within tourism, few have explored the use of edutainment within festival experiences. This oversight has created a lack in knowledge and understanding, despite the clear benefits of enjoyable learning experiences – edutainment. Moreover, it has created a gap between academia and practice; as the contributing authors have demonstrated, festivals are utilising edutainment to enhance their audience experience, yet scholars have failed to acknowledge this. In response to this oversight, the editors have assembled a carefully curated collection of chapters that include a wide range of international case studies, from science and food festivals to heritage and dark festivals. Through a variety of methodologies and methods, including interviews, observations, databases, netnography, and social media analysis in both face-to-face and digital interactions involving the festival participants, organisers, and other relevant stakeholders, the contributing authors have provided a well-rounded global perspective on how edutainment is applied within festival experiences.

This book is valuable for scholars, festival organisers, policy makers, and students interested in or studying festivals, events, edutainment, and/or experience design. Other tourism industry scholars, professionals, and students of, for example, visitor attractions, museums, theatre, and hospitality services may also find this book of value considering their established use of edutainment within their sectors.

Giulia Rossetti is a Senior Lecturer in Events Management in the Business School, Oxford Brookes University. Giulia's areas of expertise are understanding festival and event experiences using cultural sociology and serious leisure theories, festival socio-cultural impacts, and the educational value of festivals and events. Her current research interests include the role of events in generating well-being, storytelling and festivals, edutainment at festivals, intangible heritage, and cultural tourism. Giulia is the *Event Management Journal* Social Media Editor.

Brianna Wyatt is a Senior Lecturer and the current Postgraduate Subject Coordinator for the Hospitality, Tourism, and Events Management programmes at Oxford Brookes University. She specialises in dark tourism with an emphasis on interpretation and experience design, with her most recent publication on re-enactment in lighter dark tourism in *The Journal of Travel Research*. She also has industry experience working in both heritage and dark tourism, and is a Trustee and Member of the Board of Directors for the Buckingham Old Gaol. She currently sits on the Editorial Boards for the *World Leisure Journal* and *Event Management*.

Jane Ali-Knight is a Professor in Festival and Event Management at Edinburgh Napier University, Scotland, and a Visiting Professor at Curtin University, Australia. A recognised academic, she has published widely in the areas of wine tourism and regional development, destination development, festival and event marketing and management, accessibility, and well-being. Jane is a Fellow of the Royal Society of the Arts and Higher Education Academy and a trustee of British Arts Festivals Association and Without Walls: Innovators in Outdoor Arts and Hidden Door Festival.

Routledge Critical Event Studies Research Series
Editors: Rebecca Finkel, *Queen Margaret University, UK, and*
David McGillivray, *University of the West of Scotland, UK*

For more information about this series, please visit: www.routledge.com/Routledge-Critical-Event-Studies-Research-Series/book-series/RCE

Festivals and Edutainment

Edited by Giulia Rossetti, Brianna Wyatt and Jane Ali-Knight

Routledge
Taylor & Francis Group

LONDON AND NEW YORK

First published 2024
by Routledge
4 Park Square, Milton Park, Abingdon, Oxon OX14 4RN

and by Routledge
605 Third Avenue, New York, NY 10158

Routledge is an imprint of the Taylor & Francis Group, an informa business

British Library Cataloguing-in-Publication Data
A catalogue record for this book is available from the British Library

ISBN: 978-1-032-30499-1 (hbk)
ISBN: 978-1-032-30501-1 (pbk)
ISBN: 978-1-003-30541-5 (ebk)

DOI: 10.4324/9781003305415

Typeset in Times New Roman
by SPi Technologies India Pvt Ltd (Straive)

Contents

Figures

Tables

Contributors

Jane Ali-Knight, PhD, is a Professor in Festival and Event Management at Edinburgh Napier University, Scotland, and a Visiting Professor at Curtin University, Australia. A recognised academic, she has published widely in the areas of wine tourism and regional development, destination development, festival and event marketing and management, accessibility, and well-being. Jane is a Fellow of the Royal Society of the Arts and Higher Education Academy and a trustee of British Arts Festivals Association; Without Walls: Innovators in Outdoor Arts and Hidden Door Festival.

Carissa Baker is an Assistant Professor of Theme Park and Attraction Management at the University of Central Florida's Rosen College of Hospitality Management. Her interdisciplinary scholarship focuses on aspects of the themed entertainment industry, especially narrative, cultural, technological, and guest experience aspects.

Cherry Canovan is a Research Associate at the University of Central Lancashire. She obtained her PhD in Mathematical Physics from Lancaster University and previously worked as an education journalist. Her research interests include the impacts of science festivals, as well as widening participation in STEM and higher education more generally.

Cemile Ece is a PhD Candidate at Eskişehir Osmangazi University, Institute of Social Sciences, Tourism Management. She has been working as a Project Specialist for about eight years. She has done many projects in the field of tourism. Her researches focus on rural tourism, sustainability, eco-recreation, ecotourism, and grounded theory.

Efnan Ezenel is a PhD Candidate at Eskişehir Osmangazi University, Institute of Social Sciences, Tourism Management. She is a project specialist and national tourist guide. Her PhD research focuses on inclusive tourism for people with disabilities. She is also conducting research relating to accessible tourism, eco-recreation, grounded theory, and rural tourism studies.

Paul Fallon is a Senior Lecturer in the Lancashire School of Business & Enterprise at the University of Central Lancashire. In addition to teaching and working with international partners, he is an active researcher within the tourism, hospitality, and event management subject areas, with a particular interest in marketing and consumer behaviour.

Adalberto Fernandes is a researcher at the Institute of Contemporary History – NOVA University of Lisbon. He is a member of the editorial commission of *Kairos – Journal of Philosophy & Science* (De Gruyter). His research interests are science communication, science and media, science and politics, and discourse politics. His recent publications include "Science Communication as Soft Power" (in press, *The Routledge Handbook of Soft Power*, 2nd Edition), "Designing (the) Politics of Participation in Science" (accepted, *JCOM-Journal of Science Communication*), "The Appeal of Far Right Pandemic Politics: A Southern Europe Case" (2023, *Javnost-The Public*), "The Problematic Scientificy of Psychology in the Media: How Mental Illness Coverage Could Lead to Criminality Prejudice" (2022, *Tripodos*), "Communicating Corrected Risk Assessments and Uncertainty about COVID-19 in the Post-Truth era" (2021, *Frontiers in Communication*), "Science as a Virulent Myth Archive" (2020, *Social Anthropology*), and "Older peoples' Sacrifice during COVID-19 Pandemic" (2020, *Working with Older People*).

Elspeth A. Frew is an Associate Professor at La Trobe University, Melbourne. Dr Frew has 30 years of experience as a university academic, researcher, and educator in tourism and event management. Dr Frew's research interest is in cultural tourism, with a particular focus on dark tourism and industrial tourism.

Joanna Goodey is a Senior Lecturer in Events and Entertainment Management at London South Bank University. She is also an experienced events and cultural industries consultant, producer, and educator, and is passionate about access to education and sustainable development of the creative, cultural, and community sectors. Her research interests range from sustainable development and practices for the cultural, creative industries, and events sectors, through to social justice and inclusion in education and critical pedagogical practice.

Kathryn Jones is an Events Manager at the University of Central Lancashire, with key focus on Operational Events Management, Venue Management, and Large Scale Events. Building on previous industry experience, Kathryn achieved a Distinction and won awards for her MA Internship in International Tourism, Hospitality and Events Management in 2018.

Ingrid Kajzer Mitchell is an Associate Professor in Marketing at Royal Roads University. Her research areas broadly focus on marketing's impact on society, advancing sustainable business practices and purpose-driven brands, alternative production, and consumption practices with a particular interest in alternative food movements.

Nicholas Karachalis is an Assistant Professor at the Department of Tourism Economics and Management of the University of the Aegean in Chios and a researcher in the field of urban development, heritage, and cultural/ tourism strategies. He also teaches at the post-graduate programme in Cultural Management at the Hellenic Open University.

Zoe Leonard is the Research Events Coordinator within the Centre for Digital Entertainment and the Centre for Applied Creative Technologies at Bournemouth University. Zoe has an MSc in Event Marketing at Bournemouth University and has over 18 years' experience within the events industry. Zoe's passion for the 1940s era stems from her childhood; her research provides an opportunity to combine her event management skills and her passion for the 1940s era. Zoe's contribution draws on her previously unpublished MSc dissertation "A Critical Review of Authenticity, Storytelling and Nostalgia within 1940s Events in the United Kingdom, to Determine if They Present an Authentic Reflection of the Past" – 2021.

Will Low is Professor of Sustainable Business Practice at Royal Roads University. He trained as an economist at the University of British Columbia and the London School of Economics. He is internationally recognised for research which helped to define the field of fair trade studies, and he also works on labour rights and alternative food networks.

Xiao Lu (Lucia) is an Assistant Professor in the School of Cultural Technology at Xi'an Jiaotong-Liverpool University. Her research areas include the sociology of cultural consumption, arts marketing, audience development, digital engagement, and national and local policy on cultural and creative industries. Her current research project investigates the role of cultural policy in China's online game industry.

Amaia Makua worked for 14 years at the Institute of Leisure Studies at the University of Duesto, Bilbao, Spain, where she was a lecturer and researcher in leisure, tourism, and culture. Dr Makua is currently a civil servant in Leioa, the Basque Country, Spain.

Giulia Rossetti, PhD, is a Senior Lecturer in Events Management in the Business School, Oxford Brookes University. Giulia's areas of expertise are understanding festival and event experiences using cultural sociology and serious leisure theories, festival socio-cultural impacts, and the educational value of festivals and events. Her current research interests include the role of events in generating well-being, storytelling and festivals, edutainment at festivals, intangible heritage, and cultural tourism. Giulia is the *Event Management Journal* Social Media Editor.

Driselda-Patricia Sánchez-Aguirre holds a PhD in economic-management sciences and is currently a postdoctoral researcher at the Institute of Geography of the UNAM (IG), member of the national research system in Mexico, and co-leader of the Space, Culture and Tourism research group at the IG. Her research interests are cultural tourism, image, heritage, and festival management.

Elif Şenel is a PhD Candidate at Eskişehir Osmangazi University, Institute of Social Sciences, Tourism Management, and is a professional trainer as well. Her topic is tourism management and hospitality. She has researched on yoga tourism, recreation, and flow theory.

Sofoklis Skoultsos is an Assistant Professor at the Department of Economics and Sustainable Development at Harokopio University of Athens and a researcher in the field of cultural and sustainable tourism and events management. He also teaches at the undergraduate programme of Management of Tourism Enterprises at the Hellenic Open University.

Hannah Stewart, MSc, is a doctoral candidate at the University of Central Lancashire's Institute for Dark Tourism Research, specialising in Dark Event Tourism. She is presently a Production Manager with the Edinburgh Science Festival and has over a decade of international experience designing, producing, and managing festivals across Canada, the UK and the Middle East.

Louise Todd is an Associate Professor in Festival and Event Management and Deputy Lead of the Tourism Research Centre at Edinburgh Napier University, Scotland. Louise's research and teaching interests lie in arts and cultural tourism, festivals, and events, and in these settings specifically, stakeholder and community engagement, visual culture, visual research methods, art, and design.

Christine Van Winkle is a Professor in Kinesiology and Recreation Management at the University of Manitoba. Christine is committed to community-based research examining visitor experiences at events and attractions. As a former festival coordinator and attraction consultant, Dr Van Winkle brings both practical experience and theory-based research to inform practice.

Julie Whitfield is the Post Graduate Programme Leader for Events Management and is a Senior Lecturer in Conference & Events Management in the Department of Sports and Events Management at Bournemouth University Business School. Prior to joining Bournemouth University, she lectured at the Institute for Tourism Studies, Macau, China; IHTTI School of Hotel Management, Neuchâtel, Switzerland, and at Mahidol University International College, Bangkok. Julie has extensive experience of teaching Events Management to international students in a wide range of cultures and international environments. Julie is the Co-Founder of the (ICE (International Conference for Events)): Making Waves in Events Management. Julie's research predominantly focuses on the UK conference sector and the events industry.

Angela Wright, MMIIGrad, MBS, PhD, MCIPD (Academic), MATLHE, D-EduLaw, is a Senior Lecturer and researcher at Munster Technological University, Cork. Angela is Programme Director of the MBA (Strategy) at MTU and Chair of the Research Ethics Committee (REC). Angela is passionate about education and the key role it plays in career pathways. As she leads the MBA, she uses her educational voice to encourage as many learners as possible to navigate the many pathways within higher education and to reach their potential in higher-level managerial roles. Her main areas of interest include Tourism and Sustainability, Higher Education, Marketing and Business Strategy, and she holds a master's and PhD in these disciplines. A member of leading journal editorial boards, she has authored several academic books and published extensively in peer review academic journals (100+). In September 2022, she organised and chaired the International ATLAS Conference at MTU, welcoming 188 delegates from 26 countries. Angela maintains direct links to industry and is regularly asked to consult and contribute to the management and marketing issues of the day. Having served as Governor for over 15 years at Fota Wildlife Park in Cork, a major Irish tourist attraction; she continues to be a member of the Fota company, which attracts over 500,000 visitors to the region.

Brianna Wyatt, PhD, is a Senior Lecturer and the current Postgraduate Subject Coordinator for the Hospitality, Tourism, and Events Management programmes at Oxford Brookes University. She specialises in dark tourism with emphasis on interpretation and experience design, with her most recent publication on re-enactment in lighter dark tourism in *The Journal of Travel Research*. She also has industry experience working in both heritage and dark tourism, and is a Trustee and Member of the Board of Directors for the Buckingham Old Gaol. She currently sits on the Editorial Boards for the *World Leisure Journal* and *Event Management*.

Part I
Planning edutainment

1   Introduction

Brianna Wyatt and Giulia Rossetti

Introduction

Festivals are dynamic, influential, and powerful – a phenomenon that has the ability to create greater understanding of one's surroundings and subsequently social change. They exist as a space for cultural exposure that may result in what Garcia (2021, para. 10) calls "meaningful" and "life-defining shared memories". They offer meaning to varying audiences through acts of, for example, engagement, interactivity, remembrance, tradition, and feasting – all with the potential to captivate and "enable people to negotiate their sense of belonging" and/or to "(re) consider their understanding and place in the world" (Whyton, 2020, p. 556). Such experiences are made possible through strategically planned experience designs that mix educational elements with entertainment activities, thereby creating an edutainment festival experience. As such, this edited book explores the practice of edutainment, or rather the mixing of education and entertainment, within festival experiences. Through a series of case studies, it explores how a wide range of festivals utilise edutainment methods to connect with audiences as well as how edutainment is perceived by the attendees.

Edutainment

Edutainment is a varied practice, including online gaming, live-action role play, film and television, education, tourism, and events. Within the context of the leisure industry, it is traditionally understood as an interpretation agenda, designed to promote educational entertainment or entertaining education. It is commonly used to transform static and passive audience experiences to be more engaging and interactive, so that audiences may both enjoy and learn more deeply (Oren & Shani, 2012; Smith, 2016). With the aim to interpret and communicate information through engaging and entertaining activities and/or media, edutainment creates memorable audience experiences.

Etymologically, the notion of edutainment is often linked to Walt Disney, who first in 1945 and later in 1954 addressed the usefulness of mixing education and entertainment for audience learning (Oren & Shani, 2012; Wyatt et al., 2021). Although the coining of the term 'edutainment' is often attributed

DOI: 10.4324/9781003305415-2

to Robert Heyman while producing National Geographic Society documentaries in the 1970s (Aksakal, 2015; McKinney, 2018), it is often referred to as a
process, strategy, or tool for designing and improving audience experiences to
be memorable (Wyatt, 2022). Such experiences tend to use innovative and
engaging methods, such as interactive theming, storytelling, immersion, technology, simulation, and re-enactment, which others have previously noted as
beneficial for enjoyable learning (Aksakal, 2015; Hertzman et al., 2008; Rapeepisarn et al., 2006).

From a consumption point of view, the notion of edutainment underpins
Pine and Gilmore's (1998) experience economy, which theorises that experiences consist of four realms: education, entertainment, escapism, and aesthetics. In explaining the experience economy, Pine and Gilmore (1998, para. 6)
refer to the staging of experiences in such a way that they will be deeply memorable, which they explain can be accomplished "on an emotional, physical,
intellectual, or even spiritual level". They go on to address the possibility of
these experiences as being dependent on theming – a staging process similar to
set design (Åstrøm, 2020) and the engagement of senses, which is arguably very
effective in creating memorable experiences. Such claims have been echoed in
subsequent studies that have explored audience experiences reliant on edutainment techniques to create engaging and thought-provoking experiences (see
e.g. Bowman & Pezzullo, 2010; Ivanova & Light, 2018; Oren & Shani, 2012;
Wright, 2021; Wyatt et al., 2023).

The experience economy argument echoes Csikszentmihalyi's (1990) concept of flow, in which the optimal experience is both educational and fulfilling,
and one in which people are so deeply engaged that the experience becomes a
life-long learning opportunity, or rather an imprint on their life. Such experiences have been on the rise with an increasing demand for more engaging and
interactive experiences (Neuhofer & Buhalis, 2014). In fact, Neuhofer and
Buhalis (2014, p. 124) refer to the experience economy as one that has continued to evolve over the past few decades, moving from "Experience 1.0 (the
experience economy) to the Experience 2.0 (co-creation experiences), to the
now Experience 3.0 (technology enhanced experiences)". Their proposal of
this shift has rang true, as affective, co-created, immersive, simulated, and
hyper-real experiences – all made possible through edutainment methods – are
becoming a preferred experience among audiences (Alabau-Montoya &
Ruiz-Molina, 2020; Light & Ivanova, 2021; Martini & Buda, 2020). What is
more, studies have shown that the experience economy has grown to accept
edutainment as an effective way to connect with audiences on a deeper level,
engaging them in fun-centric learning experiences. Santonen and Faber (2015),
and more recently Seraphin and Yallop (2020), have demonstrated the benefit
edutainment methods can have for children and adults alike, as the application
of edutainment can generate greater learning motivation to engage and participate in learning, as well as to retain the information learned. For children in
particular, learning through play can help them to develop not only knowledge
and understanding but also their social and communication skills (Seraphin &
Yallop, 2020).

Edutainment and the festival experience

Within festival experiences, edutainment can offer added value to the audience experience. In exploring edutainment within wine festivals, Viljoen et al. (2018) demonstrated how edutainment experiences help the audience to learn something new, enhance their existing knowledge, develop a greater appreciation for local produce, diversify their palate, enhance their curiosity, and enhance their overall well-being, among a range of other benefits. More recently, Iverson et al. (2022) explored the use of edutainment activities within Viking festivals, demonstrating edutainment leads to "a deeper level of immersion", which in turn increases "the appraisal – audience attitude, memory, and intentions – of Viking festivals" (p. 37). Despite these benefits, edutainment as a subject for study within festival experiences remains underdeveloped and it also remains underutilised by festival practitioners. Some have called for more festivals to increase their edutainment opportunities in order to allow audiences not only to be entertained by festival performances but also to be educated within that space (Saifudin & Saleh, 2022; Viljoen & Kruger, 2020; Woosnam et al., 2009).

Exacerbating the underdeveloped understanding and practice of edutainment within festivals, very few have explored this topic within an academic context. Instead, edutainment and the notion of learning for fun are most often explored within museum and visitor attraction settings (e.g. Hertzman et al., 2008; Packer, 2006; Packer & Ballantyne, 2004; Seraphin & Yallop, 2020). This may be due to some suggestions that festivals are mainly occasions for entertainment and leisure (Driscoll, 2014; Ommundsen, 2009) and that, nowadays, people attend festivals mainly for fun and social purposes, and not for education (Meehan, 2005; Négrier, 2015). As such, calls for more research on how festivals can be edutaining experiences, where people learn while having fun (see e.g. Rossetti & Quinn, 2019, 2021; Szabó, 2015; Wilks & Quinn, 2016), have hitherto been unfulfilled. Given the demonstrated benefits of edutainment, there is an overt need to explore how education and entertainment mix within festivals to enhance the audience experience. Therefore, this book addresses these calls and extends understanding of how education and entertainment work together in festival contexts. As there has yet to be any texts directly relating to festivals as edutainment experiences, this book, framed thematically in four parts, offers what is believed to be the first collection of festival studies that emphasise edutainment experiences.

The curated collection of chapters includes a wide range of case studies, from science and food festivals to dark and history festivals. The case studies included are diverse in terms of size and locality – taking place across the globe in different countries, including UK, USA, Greece, Turkey, Ireland, and Mexico. A variety of methodologies and methods have also been used throughout the chapters, including interviews, observations, databases, netnography, and social media analysis in both face-to-face and digital interactions involving the festival participants, organisers, and other relevant stakeholders. All of this helps to provide a well-rounded global perspective on how festivals and edutainment are intertwined.

Part I: Planning edutainment

The first five chapters of this book explore the ways festivals plan edutainment and informal learning experiences, as well as how stakeholders can influence the designing of edutainment experiences at festivals. It is commonly understood that any good festival experience starts with effective planning for the festival and audience experience. Within this, considerations are needed for the intended audience experience and what festival attendees should be taking away from the experience, which in turn requires discussions surrounding the tools and methods most effective for achieving these intended outcomes. Evaluation is therefore of great importance in ensuring the planning processes are both completed and fulfilled. Each chapter within this first part addresses elements of the edutainment festival experience planning process, as outlined in the following.

In Chapter 2, Jones, Canovan, and Fallon explore how the organisers of the Lancashire Science Festival in the UK manage and deliver their annual edutainment experience at the University of Central Lancashire. Using a triangulation of methods, including surveys, interviews, and focus groups, their chapter highlights how a range of stakeholders work together to plan, design, deliver, and evaluate the festival to ensure both an engaging audience experience and the continuous improvement of the festival. Their findings are relevant not only for other science festivals but also for other educational institutions that stage or are considering staging edutaining festivals or student-led events.

Chapter 3 brings us to the USA, in which Stewart, Todd, and Ali-Knight examine the Haunted Happenings festival in Witch City – Salem, Massachusetts. Using a case study approach, this chapter explores how the festival uses edutainment to educate and entertain attendees about Salem's dark history of witchcraft and persecution. The authors use interviewing to reveal that the Haunted Happenings festival has transformed Salem's dark past into a festival experience that engages attendees with the local attractions through edutainment, which they argue act as a combative tool for social and political change, and as a cultural benefit that creates an immersive educational experience.

In Chapter 4, Baker examines in detail the Experimental Prototype Community of Tomorrow (EPCOT) festivals in the USA, which include the International Festival of the Arts (2017), the International Flower & Garden Festival (1994), the International Food & Wine Festival (1996), and the International Festival of the Holidays (1996). Baker outlines how each festival is different, although the festivals are similar in terms of their planned experience, interactive activities, demonstrations and educational sessions, cultural elements, and merchandise. Drawing on the author's in-person experiences and participant observation at the festivals from 2000 through to 2022, this chapter illustrates the potential of a theme park festival as revenue strategy, play site, cultural immersion creator, and educational tool.

Closing this section, in Chapter 5, Goodey explores SouthWestFest, an annual award-winning cultural community festival in the London Borough of

Westminster (UK). Drawing on a case study approach, this chapter sets out to consider how the specific cues, environment, and setting of the SouthWestFest festivalscape are utilised by the festival and three of its partner organisations, in the planning and design of learning and engagement activities which take place in connection with entertainment delivered through the festival. Goodey demonstrates the value of understanding these processes as edutainment is then assessed and presented in terms of supporting the festival in heightening learning opportunities and experiences and communicating outcomes to funders and stakeholders.

Part II: Audience engagement

Part II delves deeper into how audiences can become engaged and immersed in edutainment experiences at festivals and how edutainment can occur. In focusing on audience engagement, the four chapters of this section explore the dynamics of this important component of the festival experience. Creating engaging and interactive experiences within a festival is arguably key to ensuring audience enjoyment and satisfaction, but these are also foundational elements for edutainment experiences as well, which each of the following chapters addresses.

In Chapter 6, Lu explores how international music festivals used digital platforms to create edutainment experiences during the COVID-19 pandemic. Drawing on the data collected from online audience comments and reactions, Lu analyses the ways in which audiences obtain and extend their edutainment experiences in the 4th China-UK International Music Festival held on digital platforms. The findings of this chapter suggest the enjoyment of online music festivals stimulates audiences to learn more about Chinese and UK music and musicians, which in turn contributes to furthering their digital engagement and festival participation experiences. Importantly, this chapter prompts important questions for the future of festivals online as most festivals are now being delivered again in the physical world. Still, there are benefits of offering online or hybrid experiences, which certainly allow for greater access and inclusivity.

Exploring the British Science Festival in Chapter 7, Frew and Makua address how science festivals have become a popular means for the science community. They outline how these festivals help to raise public awareness, as well as encouraging conversations about science between science communities and society. The authors take a unique approach as they consider science festivals within the theory of valuable leisure, which suggests successful leisure activities (such as science festivals) can contribute to human development by enhancing individuals' capabilities, promoting well-being, empowering them, encouraging conviviality, supporting individual values, and protecting cultural diversity. Their findings reveal that creative and innovative activities and experiences via edutainment is an approach that is aligned with the key aspects of the valuable leisure experience.

Chapter 8 explores seven Greek festivals in which the authors, Skoultsos and Karachalis, reveal these festival experiences create edutainment opportunities that can cultivate strong bonds between the local community, tourists, and festival organisations. As the authors approach edutainment as a tool to enhance the active engagement of attendees during the festival experience, their findings, collected from structured interviews, reveal edutainment activities help to develop a form of festival legacy that further contributes to festival success and long-term viability.

Part III: Sustainability and EDI

Part III of this book explores the nature of sustainability and EDI (equality, diversity, and inclusion) within the context of edutainment activities and festival experiences. Considering edutainment is often used to engage individuals in their experience and with others in a way that allows them to co-create their experience so that it is personalised and deeply memorable, ensuring these experiences are equally accessible, inclusive, and promote diversity is of the utmost importance. Such assurances help to ensure the festival experience is a sustainable one – not just in terms of being green and environmentally friendly – but also socially and economically sustainable, thereby extending the festival's legacy value. The three chapters within this part highlight important details about how edutainment is an essential tool for festival experiences that seek to promote these themes.

In Chapter 9, Wright explores how the rural farming Irish festival – The Winterage Festival – fosters learning about nature, farming, and the locality through edutainment techniques that promote place-based learning centred around heritage, folklore, sustainability, and biodiversity. Drawing on interviews and participation in the visitor journey, the author's findings suggest that education and entertainment coincide in this lunar landscape where reverse transhumance is practised – moving cattle to upland pastures over the winter months, before returning to fertile lowland pastures for the summer. Most insightful is the author's revelations that visitors and locals alike are fully immersed in this festival, where co-creation leads to the learning landscape providing the canvas for an outdoor classroom experience.

The Accessible Film Festival, an annual festival held in Turkey, is explored in Chapter 10. The authors Ece, Ezenel, and Şenel consider the issues of festival organisation, transportation and location, physical environment, social inclusion, communication, content, and technology throughout this comprehensive chapter in order to ensure inclusiveness and diversity in an edutainment-driven festival. Relying on interviewing and storytelling techniques, the authors demonstrate how festivals can utilise edutainment methods to create film festival experiences that are inclusive and diverse for both attendees and filmmakers showcasing their work at film festivals.

In Chapter 11, Sánchez-Aguirre explores film festivals as a way to enhance sustainability understanding through edutainment techniques. Through a case study approach reliant on interviews and media analysis, the author explores Cinema Planeta, the first film festival in Morelos, Mexico, as an example of how edutainment techniques can be used to engage film festival audiences through online platforms. The author's findings highlighted the significance of including an online platform within the film festival experience, as well as giving audience rewards to those for participating in edutainment activities relating to the themes of the films. In doing so, the author demonstrates how edutainment activities may be designed to promote sustainability and the United Nations Sustainable Development Goals (2015) through film festival experiences.

Part IV: Experiencing edutainment

The final part of this book focuses on understanding how festival attendees experience edutainment, and not just enjoyable learning but also their development of a sense of authenticity and community. The festival experience has, up until now, been discussed throughout this book in a variety of contexts relating to the planning and organisation processes for ensuring an enhanced, inclusive, and sustainable audience experience through edutainment activities. The final three chapters therefore explore edutainment within the festival experiences, with a closer look into how audiences experience enjoyable learning – edutainment.

In Chapter 12, Leonard and Whitfield explore 1940s festivals and events in the UK, which are shown to create an immersive and edutaining experience that is guided by homage and commemoration of this time. The authors complete a critical questioning on the importance of storytelling through living history to disseminate memoirs of this historic era, how 1940s festivals and events can trigger nostalgia, the inclusion of edutainment and finding a synergy between entertainment and education, and the responsibility to accurately present this past as best possible. Their findings confirm that organisers strive to present an accurate and authentic representation of the 1940s in order to help attendees to better connect with and learn from this past. Attendees are thus engaged in the experience through truthful storytelling and entertaining festival content, which work together to form a successful edutainment experience.

In Chapter 13, Kajzer Mitchell, Van Winkle, and Low merge two concepts – ludic learning spaces and service ecosystems – to examine how two food festivals, the globally occurring Local Wild Food Challenge and the Harvest Moon Festival in Canada, engage their attendees in deep learning processes. The authors use this framework to examine interactions between multiple actors and to map the resulting multi-layered edutainment experiences through various levels of analysis, from individual to relational, and from personal to communal. Their findings reveal how edutainment festival experiences do not just constitute a fun day out but that the fun and frolics within the festival encourage attendees to explore their creativity, express themselves in novel ways, learn about alternative food realities, and connect with a broader alternative food movement.

Chapter 14 is a theoretical contribution that proposes an enlarged concept of emotion for edutainment in science festivals. In this chapter, Fernandes explores how science festivals that use edutainment techniques can capitalise on negative emotions to engage attendees that have a troubled relationship with science. Using specifically Horror Science-Fiction Movie Festivals, the chapter questions the theory that edutainment-based science festivals are appealing because they rest precisely on positive emotions. As such, the author posits negative emotions can be just as powerful as positive emotions when used effectively through edutainment techniques in science-themed festivals and can in turn generate an enjoyable learning experience.

Conclusions

This book closes with an afterword (Chapter 15), in which the editors review the ideation of this book. A reflective review of the contributing chapters is offered before addressing the future of edutainment within the festival experiencescape. Themes such as technology and well-being are identified as growing themes for academic interest across interdisciplinary studies relating to the global visitor economy, and specifically the festival industry. Given these future thinking proposals, in conjunction with the following carefully curated collection of chapters, this book will be a valuable tool for students, researchers, festival organisers, and policy makers in better understanding the edutainment potential of festivals. Moreover, it will be useful for people working in higher education and involved in pedagogic research focused on using festivals as a tool for enjoyable learning opportunities. Other fields that utilise edutainment to create audience experiences, such as museums, hospitality, theatre, and education will also find interest in this book and the authors' contributing chapters.

References

Aksakal, N. (2015). Theoretical view to the approach of the edutainment. *Procedia – Social and Behavioral Sciences, 186*, 1232–1239. https://doi.org/10.1016/j.sbspro.2015. 04.081

Alabau-Montoya, J., & Ruiz-Molina, M. E. (2020). Enhancing visitor experience with war heritage tourism through information and communication technologies: Evidence from Spanish Civil War museums and sites. *Journal of Heritage Tourism, 15*(5), 500–510. https://doi.org/10.1080/1743873X.2019.1692853

Åstrøm, J. K. (2020). Why theming? Identifying the purposes of theming in tourism. *Journal of Quality Assurance in Hospitality & Tourism, 21*(3), 245–266. https://doi. org/10.1080/1528008X.2019.1658147

Bowman, M. S., & Pezzullo, P. C. (2010). What's so dark about dark tourism?: Death, tours, and performance. *Tourist Studies, 9*(3), 187–202. https://doi.org/10.1177% 2F1468797610382699

Csikszentmihalyi, M. (1990). *Flow. The psychology of optimal experience* (pp. 1–22). Harper & Row.

Disney, W. (1945). Mickey as professor. *Public Opinion Quarterly, 9*(2), 119–125. https://doi.org/10.1086/265726

Disney, W. (1954). Educational values in factual nature pictures. *Phi Delta Kappa International, 33*(2), 82–84.

Driscoll, B. (2014). The middlebrow pleasures of literary festivals. In B. Driscoll (Eds.), *The new literary middlebrow* (pp. 152–193). Palgrave Macmillan UK.

Garcia, B. (2021, 2 Aug). *Why festivals and special events matter now more than ever.* UKRI. https://www.ukri.org/blog/why-festivals-and-special-events-matter-now-more-than-ever/

Hertzman, E., Anderson, D., & Rowley, S. (2008). Edutainment heritage tourist attractions: A portrait of visitors' experiences at Storyeum. *Museum Management and Curatorship, 23*(2), 155–175. https://doi.org/10.1080/09647770802012227

Ivanova, P., & Light, D. (2018). 'It's not that we like death or anything': Exploring the motivations and experiences of visitors to a lighter dark tourism attraction. *Journal of Heritage Tourism, 13*(4), 356–369. https://doi.org/10.1080/1743873X.2017.1371181

Iverson, N. M., Foley, C., & Hem, L. E. (2022). The role of immersive festival experiences, identity, and memory in cultural heritage tourism. *Event Management, 27*(1), 33–50. https://doi.org/10.3727/152599522X16419948694847

Light, D., & Ivanova, P. (2021). Thanatopsis and mortality mediation within "lightest" dark tourism. *Tourism Review, 77*(2), 622–635. https://doi.org/10.1108/TR-03-2021-0106

Martini, A., & Buda, D. M. (2020). Dark tourism and affect: Framing places of death and disaster. *Current Issues in Tourism, 23*(6), 679–692. https://doi.org/10.1080/13683500.2018.1518972

McKinney, S. (2018) *Educational techniques and methodology.* Ed-tech Press.

Meehan, M. (2005). The word made flesh: Festival, carnality, and literary consumption. *Text* (Special Issue n.4).

Négrier, E. (2015). Festivalisation: Patterns and limits. In C. Newbold, C. Maughan, J. Jordan, & F. Bianchini (Eds.), *Focus on festivals: Contemporary European case studies and perspectives* (1st ed., pp. 18–27). Goodfellow Publishers Limited.

Neuhofer, B., & Buhalis, D. (2014). Experience, co-creation and technology: Issues, challenges and trends for technology enhanced tourism experiences. In S. McCabe (Ed.), *The Routledge handbook of tourism marketing* (pp. 124–139). Routledge.

Ommundsen, W. (2009). Literary festivals and cultural consumption. *Australian Literary Studies, 24*(1), 19–34.

Oren, G., & Shani, A. (2012). The Yad Vashem Holocaust Museum: Educational dark tourism in a futuristic form. *Journal of Heritage Tourism, 7*(3), 255–270. https://doi.org/10.1080/1743873X.2012.701630

Packer, J. (2006). Learning for fun: The unique contribution of educational leisure experiences. *Curator: The Museum Journal, 49*(3), 329–344.

Packer, J., & Ballantyne, R. (2004). Is educational leisure a contradiction in terms? Exploring the synergy of education and entertainment. *Annals of Leisure Research, 7*, 54–71. https://doi.org/10.1080/11745398.2004.10600939

Pine II, B. J., & Gilmore, J. H. (1998). Welcome to the experience economy. *Harvard Business Review*, 97–105. https://hbr.org/1998/07/welcome-to-the-experience-economy

Rapeepisarn, K., Wong, K. W., Fung, C. C., & Depickere, A. (2006). Similarities and differences between "learn through play" and "edutainment". In *Proceedings of the 3rd Australasian Conference on Interactive Entertainment*, 4–6 December 2006, Perth, WA, pp. 28–32.

Rossetti, G., & Quinn, B. (2019). Insights into the learning dimensions of literary festival experiences In I. Jenkins & A. K. Lund (Eds.), *Literary tourism: Theories, practice and case studies* (pp. 93–105). CABI.

Rossetti, G., & Quinn, B. (2021). Understanding the cultural potential of rural festivals: A conceptual framework of cultural capital development. *Journal of Rural Studies, 86*, 46–53. https://doi.org/10.1016/j.jrurstud.2021.05.009

Saifudin, M., & Saleh, M. (2022). Edutainment in sustainability: Can sustainability communication also be fun? In M. Saifudin, M. Saleh, N. A. A. Rahman, & S. A. A. Kasuma (Eds.), *Sustainability communication across Asia: Fundamental principles, digital strategies and community engagement* (pp. 188–199). Routledge.

Santonen, T., & Faber, E. (2015). Towards a comprehensive framework to analyse edutainment applications. *ISPIM Conference Proceedings, 358*(June), 1–11.

Seraphin, H., & Yallop, A. (2020). An analysis of children's play in resort mini-clubs: Potential strategic implications for the hospitality and tourism industry. *World Leisure Journal, 62*(2), 114–131. https://doi.org/10.1080/16078055.2019.1669216

Smith, M. (2016). *Issues in cultural tourism studies* (3rd ed.). Routledge.

Szabó, J. Z. (2015). Festivals, conformity and socialisation. In C. Newbold, C. Maugan, J. Jordan, & B. Franco (Eds.), *Focus on festivals. Contemporary European case studies and perspectives* (pp. 40–52). Goodfellow Publishers Limited.

United Nation Sustainable Development Goals. (2015). https://www.un.org/sustainable development/

Viljoen, A., & Kruger, M. (2020). The "art" of creative food experiences: A dimension-based typology. *International Journal of Gastronomy and Food Science, 21*, 100239. https://doi.org/10.1016/j.ijgfs.2020.100239

Viljoen, A., Kruger, M., & Saayman, M. (2018). The art of tastings: Enhancing the arts festival experience. *International Journal of Event and Festival Management, 9*(3), 246–265. https://doi.org/10.1108/IJEFM-03-2018-0023

Whyton, T. (2020). Space is the place: European jazz festivals as cultural heritage sites. *International Journal of Heritage Studies, 26*(6), 547–557. https://doi.org/10.1080/13527258.2018.1517375

Wilks, L., & Quinn, B. (2016). Linking social capital, cultural capital and heterotopia at the folk festival. *Journal of Comparative Research in Anthropology and Sociology, 7*(1), 23–39.

Woosnam, K. M., McElroy, K. E., & Van Winkle, C. M. (2009). the role of personal values in determining tourist motivations: An application to the Winnipeg fringe theatre festival, a cultural special event. *Journal of Hospitality Marketing & Management, 18*(5), 500–511. https://doi.org/10.1080/19368620902950071

Wright, D. W. M. (2021). Immersive dark tourism experiences: Storytelling at dark tourism attractions in the age of 'the immersive death'. In M. H. Jacobsen (Ed.), *The age of spectacular death* (pp. 89–109). Routledge.

Wyatt, B. (2022). Edutainment. In D. Buhalis (Ed.), *Encyclopedia of tourism management and marketing*. Edward Elgar Publishing.

Wyatt, B., Leask, A., & Barron, P. (2021). Designing dark tourism experiences: An exploration of edutainment interpretation at lighter dark visitor attractions. *Journal of Heritage Tourism, 16*(4), 433–449. https://doi.org/10.1080/1743873X.2020.1858087

Wyatt, B., Leask, A., & Barron, P. (2023). Re-enactment in lighter dark tourism: An exploration of re-enactor tour guides and their perspectives on creating visitor experiences. *Journal of Travel Research*, OnlineFirst. https://doi.org/10.1177/00472875221151074

2 Science Non-Friction

Balancing Operational and Legacy Agendas for a University-Based Science Festival

Kathryn Jones, Cherry Canovan and Paul Fallon

Introduction

Science is fundamental to our understanding of where we have come from, where we are now, where we are going, and how we improve our lives. Society was heavily reminded of its importance via the frequent call to 'follow the science' during the height of the Covid-19 pandemic (Colman et al., 2021, p. 2). Despite this significance, there is longstanding concern about a "swing away" from science (Osborne et al., 2003, p. 1050). Van Griethuijsen et al. (2015) highlight that interest in science among primary and secondary school students in Western European countries is "low and seems to be decreasing" (p. 581). Governments and other interested organisations have sought to address this perceived shortfall using interventions via both the school curriculum and less formal means. It is widely recognised that the more opportunities available for children to ask questions, make observations, and learn through hands-on experiences, the more easily they learn and connect to the real world (Erduran et al., 2006; Halat, 2007). It is therefore unsurprising that there has been a significant global growth in events such as science festivals (Canovan, 2019), which aim to appeal to a broad spectrum of the population including both children and adults, with at least 60 taking place annually in the UK alone (Kerr, 2020).

Science festivals represent "highly influential science communication events [which are] 'very much at the hub' of informal science learning and public engagement initiatives" (Peterman et al., 2020, p. 205). In the UK, these festivals are seen as an intervention strategy to increase both public engagement in science and, more specifically, levels of participation in post-16 STEM (science, technology, engineering, and maths) education. Science festivals are likely to be particularly appealing to children because younger people, particularly those of primary age, overwhelmingly express an interest in science. The large-scale UK ASPIRES project (Archer et al., 2013) found that nearly three-quarters of the English ten-year-olds surveyed said they learned interesting things in science. Moreover, this positivity is not restricted to more affluent groups; research conducted in one of the UK's most deprived areas found the children there were every bit as interested in science as their peers nationally (Canovan &

DOI: 10.4324/9781003305415-3

Walsh, 2020). Despite these drivers, staging science festivals represents a multi-faceted challenge. From an operational perspective, the eventscape and experience are required to provide the right balance between 'edutainment' (and often 'wow') and basic factors such as accessibility and safety. Furthermore, the festival is expected to deliver an intervention which creates the desired legacy in terms of visitors' science interest and attitudes. Drawing on these considerations, this chapter adopts a case study approach to explore how the Lancashire Science Festival, held annually at the University of Central Lancashire (UCLan), is designed and evaluated to meet these challenges. Held over school days and the weekend, the festival incorporates both educational and fun experiences for its target audiences of both school and family groups.

Science Festivals – A Perfect Match

The science festival proposition is soaring in popularity in terms of events and visitor numbers. Bultitude (2014, p. 175) found a "marked acceleration in the rate at which science festivals are being set up" in the first decade of the 21st century, while the National Science Foundation has highlighted a "remarkable growth in science festival activity" in the USA (National Science Foundation, 2016).

Although there is no commonly accepted typology of festivals and the term is often misapplied and commercialised, festivals share some common characteristics. Festivals represent themed celebrations with wide-ranging roles and impacts for multiple stakeholders, including for the social fabric and cultural capital of the communities which host them (Zhang et al., 2019). Additionally, festivals create experiences (Gannon et al., 2019) – they "premise their very existence on interaction and the exchange of flows (of people, information, ideas, money, cultural expressions etc." (Quinn, 2006, p. 301). Taken as a whole, the festival experience is seen as rich, interactive, immersive, and memorable. Consequently, the incorporation of the word 'festival' into an event title potentially puts a different spin on it, with the significant consequence that the event then needs to live up to its 'festival' billing.

Within the specific context of science festivals, such an experience has the potential to build what is known as 'science capital' in its audience, particularly among the younger members. Archer et al. (2016, p. 2) explain that:

> The concept of science capital can be imagined like a 'holdall', or bag, containing all the science-related knowledge, attitudes, experiences and resources that you acquire through life. It includes what science you know, how you think about science (your attitudes and dispositions), who you know (e.g., if your parents are very interested in science) and what sort of everyday engagement you have with science.

Seen in this context, the science festival appeal is twofold. First, it has the capacity to positively alter attitudes towards science, build knowledge, and add to the 'holdall', thus making it very attractive to those who aim to build science capital in the community. Given the concern about engagement with science especially among younger people and that knowledge can be disseminated very effectively through informal channels (Quinn, 2006), it is unsurprising that science festivals have grown in popularity. Second, and notwithstanding that it is important for reasons for equity to reach beyond this group, a ready-made audience comprising those who already possess high science capital and who will relish the opportunity to explore this interest in a fun and exciting environment already exists.

A further critical success factor for science festivals relates to timing. Festivals can run over a range of time periods and offer 'sacred' and/or 'profane' time experiences (Jepson & Clarke, 2014). Key factors here relate to both demand and supply, especially related to availability of venue, resources, exhibitors/performers, and attendees. The decision when to hold them can affect the atmosphere created, in terms of both who is attending and their mood at the time. Time-based factors also include pre-, peri-, and post-consumption stages (Brown et al., 2019), which contribute in different ways to the overall experience. For example, attendees progress from anticipation prior to the event to active engagement and activity sharing, which extends beyond the event and creates anticipation for the next one (Packer & Ballantyne, 2011). It is logical to assume that the more captivating, intense, and exciting the experience (Cabanas, 2020), the greater the sense of anticipation will be. Furthermore, depending on the event duration, attendees may return or extend their visit to the current one.

This temporality also allows for a focused effort on public engagement from a wide range of scientific professionals, which provides a qualitatively different experience from other informal educational experiences, such as those provided by science museums and centres. Jensen and Buckley (2014, p. 560) note that

> … investment may be made in a level of activity which would be hard to sustain for a longer period. For example, many science festivals also have high levels of intensive volunteer participation by scientists, university students, technologists and engineers…. The number and scope of the involvement of this volunteer scientific expertise plays a key role in positive visitor impacts.

The authors additionally point out that the temporary nature of the festival means that a greater focus can be placed on current research issues.

The Case of Lancashire Science Festival (LSF)

The LSF is an annual event run at UCLan in Preston, in North West England, during the early summer. The three-day festival takes over the city-based campus (see Figure 2.1), traditionally welcoming 4,000–6,000 primary school children across two school days and 6,000–8,000 attendees on a public open day.

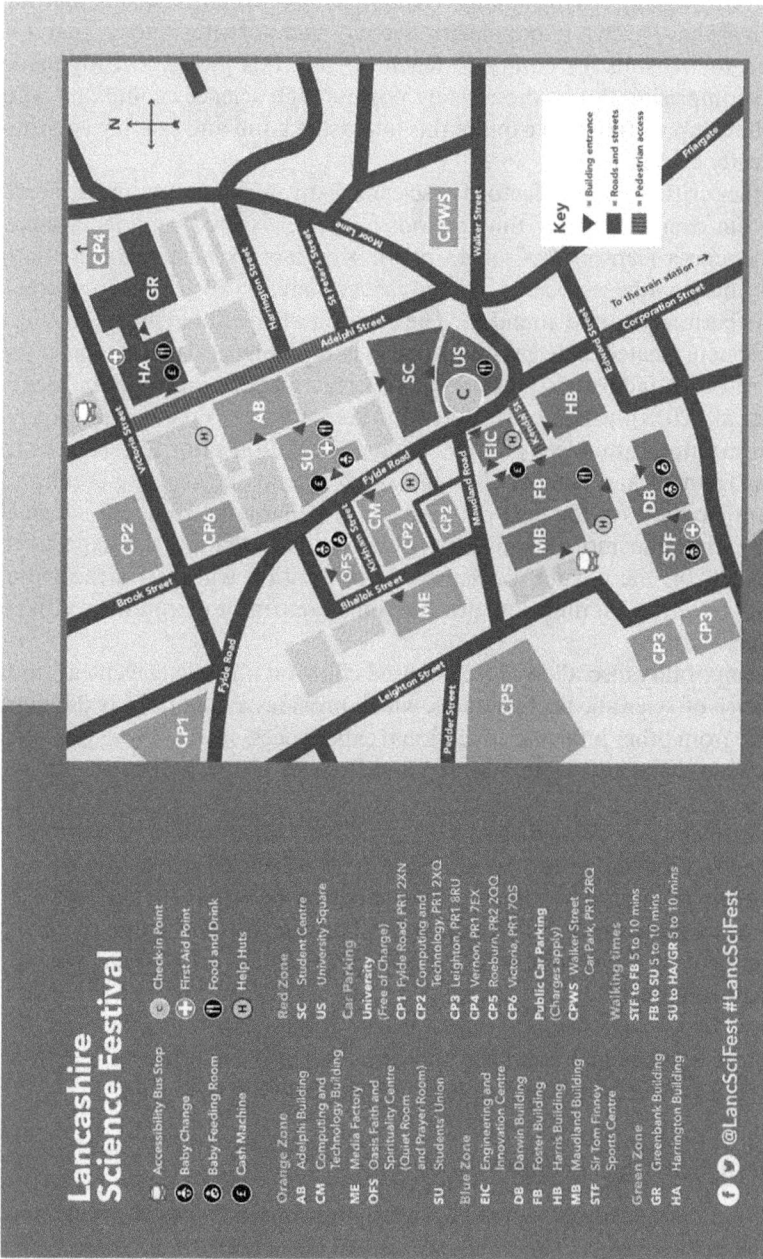

Figure 2.1 LSF map. October 2021.
Copyright University of Central Lancashire.

Thursday and Friday are dedicated to school parties, while Saturday is aimed at family groups, particularly those with primary-age children.

The purpose of the festival is to engage, inspire, and educate, focusing on widening participation, the promotion of STEM subjects and the delivery of a safe event. To do this well, the festival must provide an edutainment experience that is effectively operationally managed. Two hundred and thirty scientists and 30 organisations typically support the creation of this experience via four broad types of interpretative activity: shows/demonstrations (held in lecture theatres), pre-bookable workshops, drop-in sessions, walk-around performances, and two 'market' areas (with stands featuring practical activities).

With some justification, the university presents LSF as an 'extravaganza', and in 2021 it "feature[d] a packed programme of jaw-dropping live shows, eye-popping demonstrations and hands-on workshops led by some of today's leading lights in the STEM fields" (University of Central Lancashire, 2021). LSF has won the Best Community or Business Engagement Campaign prize at the Heist Awards for Education Marketing and Best Corporate Social Responsibility Campaign at the Northern Marketing Awards.

Methodology

This chapter explores edutainment via a case study relating to the LSF. It draws on the experience of both the University Events Team and specialist evaluators to give a multi-faceted picture of the festival. It particularly focuses on its university setting, which both provides specific operational considerations and brings with it a public duty to widen participation in science and higher education more generally.

The LSF has been informally evaluated since its inception in 2012, and since 2017 more formal methods have been used by a specialist in-house team of academic researchers. For each iteration of the festival, a range of demographic information is gathered about all visitors, including level of education, science knowledge, and postcode as a proxy for socioeconomic status. Participants are surveyed after the event in terms of their enjoyment, their learning experience, and operational matters. Edutainment, viewed as the nexus between enjoyment and education, is a key aspect of this evaluation; as the researchers note, "[f]inding an event fun and/or interesting is an important condition for gaining benefit from it" (Canovan, 2019, p. 10).

Several formal research projects acknowledging *inter alia* views of science, factual and affective learning, and audience diversification have been conducted by the in-house research team (Canovan 2019, 2020a, 2020b), utilising methods such as online surveys, on-the-day interviews, telephone interviews, and focus groups. These projects, and their related methods, are all approved by the UCLan ethics committee. Results cited in this chapter have all been gathered during such studies.

As well as drawing on the extant LSF research base, this chapter also incorporates the lived experience of the organising team and internal operational material – both of which the authors had access to – and thus provides a triangulation of data sources. For the event organisers, active interviews were used to gather data; active interviews represent a process in which both researchers and participants contribute to, therefore co-create, the discussion (Fallon & Robinson, 2017). The additional sources illustrated how both research and effective operational management are essential in the delivery of a successful edutainment experience. For example, following feedback from the previous event and learnings identified by the Events Team, an event action plan is created and updated throughout the planning process of the current festival. This plan subsequently leads to the creation of other documentation, including an event budget, site maps and floorplans, staffing plans, briefing documents, event build and break schedule, risk assessments, and event safety management plans. The organising team's wealth of experience in delivering other large events on our Preston Campus, including Graduation events, Open Days, VIP events, and School Visit Days also provides valuable learning opportunities.

Results and Discussion

For the LSF organising team – and given that the campus venue functions in its normal University role for the festival duration – working with temporal, spatial, and people dimensions is the key to a successful and harmonious event. Preparation is crucial, and the team is already in the cycle of anticipating the next incarnation of the event as the existing one is coming to fruition (Packer & Ballantyne, 2011). Furthermore, in contrast to many events where evaluation is primarily operational and/or 'one off', the evaluation function and research team are an integral part of LSF's planning and design.

LSF People

The success of any event is heavily dependent upon its people, their interconnectivity, and flows of information (Quinn, 2006). Following Home Office (2021) guidance on critical incident management, staffing for LSF includes operational (Bronze) level managers running the festival, a tactical (Silver) manager dealing with critical incidents, an Emergency Management Specialist, and a strategic (Gold) manager on standby. School guides, check-in staff, way finders, security, first aiders, and cleansing and catering colleagues cover operational duties. With the festival budget coming from university funds and external sponsorship, it is essential to achieve quality experiences at minimal cost (Tum et al., 2006). This is achieved through recruiting University colleagues to fill key positions and involving relevant internal teams in operational work.

LSF offers highly engaging but potentially hazardous activities – including chemical displays, pyrotechnics and flying rockets, robots and performers, petting zoos, and building activities – which make the festival what it is, therefore requiring a specialist safety management staffing resource.

Space and Time

Festival experiences are created within a physical and social environment. Car-neiroa et al. (2019) reflect on the importance of the venue facilities, location, accessibility, information, catering, staff, and entertainment in contributing to attendee satisfaction. Additional infrastructure such as walkways, signage, and queuing systems provide a "designed environment" to improve experiences (Berridge, 2011, p. 75). To facilitate a successful edutainment experience, the campus requires adaptation to ensure its suitability and flexibility at different stages of the experience (Brown et al., 2019). On the school days, the largest car park on site is adapted to become a coach park, including barriered walkways for staff and school groups. A large space is required for check-in on the public day to ensure sufficient space for safe queuing and timely check-in of guests. In 2021, University Square (UCLan's new outdoor venue) was utilised for this, with the team bringing in a marquee, furniture, electricity infrastructure, and barriers. The University Sports Hall was transformed into a 'Science Show-floor' exhibition space, requiring carpeting. Lecture theatres typically become show theatres and are adapted individually for each act, with queuing systems put in place to manage safe access. Signage is put up across campus to support attendees in accessing activities. As the venue remains a working university for the duration of LSF, there is a further fine balancing act in providing for the needs of event attendees, while also ensuring that students, staff, and other stakeholders are satisfied.

McLoughlin (2015, p. 246) states that event organisers must also "design and create an event which provides the right level of challenge or stimulation to the skill set of the target audience". The organising team recognises the need to 'pitch' and 'pivot' LSF to appeal to different age groups and meet different agendas (Berridge, 2011). For example, a carefully managed pre-event communications plan increases awareness, builds excitement, and especially encourages locals (including widening participation areas) to attend. It is designed to set high but realistic expectations among different audiences of primary school-age children, teachers, and families.

Festival Evaluation

Evaluation is vitally important for festivals, given their utilisation of resources, increasing need for accountability, and as a process for reflection and review (Allen et al., 2008). Science festival evaluation poses a unique set of challenges. The science festival can be evaluated from a wide range of operational and 'legacy' perspectives, including impact on parental attitudes to science (Canovan, 2019), preferences for the format of engagements (Fogg-Rogers et al., 2015), factual and attitudinal learning (Canovan, 2020a), and understanding of complex ethical issues (Rose et al., 2017).

The fact that the LSF is hosted by UCLan adds an extra dimension to the evaluation effort. The festival forms part of the institution's Access and

Participation Plan, a document stating how it will improve equality of opportunity in higher education and which must be approved by the Office for Students. In order to meet its aim of highlighting opportunities in science to visitors, it is crucial that rigorous evaluation is conducted. Consequently, UCLan employs a dedicated team of academic researchers focusing on the deeper impacts of such outreach activities. LSF is therefore assessed for its effectiveness, customer satisfaction, and on its educational impacts in terms of learning and attitudes to science. The research team uses a variety of study techniques, gathering qualitative and quantitative data both before and after the festival, and sometimes on the day as well. While mainly focusing on learning and societal impacts, operational evaluation is also conducted.

The demographic profile of science festival visitors represents a particularly pressing issue, especially given the concern that audiences tend to be composed of well-educated individuals, with a lack of ethnic and/or socioeconomic diversity (Bultitude, 2014; Kennedy et al., 2017; Manning et al., 2013). Kennedy et al. (2017, p.15) give a strong call to action to the sector:

> [Science festivals] are disproportionately reaching economically privileged and educated audiences already invested in science, as opposed to diverse and broadly representative samples of the general public.... There is a clear need for improved practices and on-going evaluation to ensure science festivals include those who are not already scientifically converted.

One of LSF's aims is to widen participation in science among underrepresented communities, resulting in a particular focus on assessing potential beneficial effects on attendees from lower-socioeconomic status backgrounds. It is consistently the case that, as with other similar events, LSF visitors are more highly educated and affluent than the local and general population. Conducting analysis on this point has allowed LSF organisers to trial schemes to improve the balance of attendees, for example, by offering a 'Community Pass' with benefits such as free meals (Canovan, 2020b). This illustrates the value to festivals of ensuring that their evaluation includes demographic analysis.

Turning to the festival's impact, science festival organisers' primary objective in staging their events is to change attitudes to science, for example, to inspire excitement or to raise awareness of the importance and breadth of science (Canovan, 2020a). LSF researchers evaluate its impact in terms of both enjoyment and learning. Studies in previous years (Canovan, 2019) have found clear evidence of participants having fun at the event via comments such as these: "It's fun, not nerdy"; "Makes science much more exciting, it's fascinating"; and "It's fantastic.... I find it exhilaratingly informative".

Changes in behaviour have been noticed by parents very quickly: "My 7-year-old came straight home and got all his science kits out of the cupboard and spent the rest of the weekend creating experiments"; "The UCLan science festival has ignited a passion for my children playing 'scientists' – start them off young! Potions are the current favourite"; and "My children, they kind of went round telling everybody, even people on the street, where they'd been, and they talked about it for weeks". It is also common for school children, following their school visit, to 'influence' their parents and families to return at the weekend.

Canovan's (2019) study also considered the learning impacts on parents accompanying their children, with 70% reporting a positive impact on their perception of science. This finding is important from a widening participation perspective, because parental interest in science correlates with post-compulsory science participation for underrepresented groups (DeWitt et al., 2016).

While learning in the affective domain is the key goal, event organisers also understand the necessity of underpinning this with cognitive learning, in order to give a realistic impression of the scientific process and avoid presenting this in an "easy and unproblematic" way (Eshach 2007, p. 172). LSF researchers have conducted extensive studies using various indicators to assess how well these learning goals are met. In one project conducted with pupils aged 9–11 (Canovan, 2020a), 95% of children agreed that the science at the festival was interesting, and 45% were able to state a science fact learned at the festival, suggesting a positive factual learning impact. The breadth of learning that occurs at the festival is demonstrated by evaluation data from the 2021 LSF and published here for the first time. Adults were asked to '[t]ell us a science fact that you or your child learned at the festival', with 75% of participants stating a piece of factual learning. Although learning topics are skewed to LSF activities, these are wide in scope, covering physical science, the natural world, human biology, and more. Examples of responses include "A day on Venus lasts longer than a year!!"; "Bubbles always become spheres regardless of what shape they start"; and "The part of the brain that controls the eyes is at the back, not behind the eyes".

However, another key goal of organisers – raising awareness of science careers – is proving more challenging to achieve. In the pupil survey cited above, although 42% of young respondents said they had learned about new science jobs, only 20% could name one. The picture was more positive among parents; many participants commented that they had learned about the breadth of science jobs available, with a majority (52%) feeling more positive about their child pursuing a career in science after attending (Canovan, 2019).

LSF constantly adapts in response to evaluation of findings, and a greater emphasis is now placed on career learning. This focus has been designed into the eventscape using both visible and invisible mechanisms (Ferdinand & Williams, 2018), including stallholders wearing 'I am a scientist' badges and presenters explaining their own progression. Although organisers have not yet tested this new approach among children, it seems to be effective for adults. The 2021 evaluation asked parents to '[p]lease name a science career that you learned about during your visit', and nearly 80% were able to name a science job. While it will be important in future to assess whether this career learning extends from parents to children, and although not the main target audience, the capacity for influencing parents in this way represents a successful outcome as they are then better able to share advice about future choices with their children.

Conclusion

Hosting the LSF at UCLan represents a virtuous circle for its various stake-holders, including the academics who deliver activities, external contributors such as science communication professionals and charities, sponsors, and local community members and schools who attend the festival. To summarise, key points include the following:

- Although not purposely designed for festivals and with its 'normal' activities running simultaneously, the university campus provides a modern, multi-dimensional, and flexible eventscape which is suitable for different science festival audiences.
- The university context offers an 'out of the ordinary' edutainment experience for its audiences and participants while providing a complimentary and contemporary setting for the festival theme and its various aims, including its research agenda.
- LSF harnesses the expertise of various groups within UCLan, including its Events Team, STEM subject specialists, the Health & Safety team, and specialist researchers, which combine to develop and deliver a safe and edutaining experience.
- LSF contributes to UCLan's commitment to national and institutional widening participation agendas via a high profile, memorable, and 'fact and fun-filled' event which builds science capital for different groups within the local community.
- LSF positions UCLan in the minds of the community as an exciting place for learning.
- The involvement of an in-house research team enables rigorous and multi-dimensional evaluation, which allows for relevant changes and innovations to the festival, as well as research publications; research tools are especially attuned to different audience groups and the affective-cognitive learning continuum.
- Finally, LSF transforms the campus and creates a unique focus and 'buzz' among its organisers, demonstrators, volunteers, and attendees; although there can be tensions with other campus user groups, it energises the campus at a currently less busy time of year.

The foundations for LSF's ongoing success lie in its people and processes, especially those related to its design and evaluation, and the balancing acts which they facilitate across a range of dimensions. Given the importance of science for mankind, it is hoped that these are positive messages for other universities and science festival organisers in terms of how they can combine to support and monitor the development of 'science capital'. Furthermore, the principles and practices presented here can be helpful to, and be developed further by, the organisers of other types of festivals staged in educational and university settings, and especially those whose legacy relates to changes in our communities' awareness and behaviours. From an organisational perspective, involvement in edutainment events such as LSF may be beneficial in the development of local, regional, and national profiles and provide competitive advantage as a medium- to long-term investment in terms of student recruitment.

References

Allen, J., O'Toole, W., Harris, R., & McDonnell, I. (2008). *Festival and special event management* (4th ed.). John Wiley & Sons.

Archer, L., Dawson, E., DeWitt, J., Godec, S., King, H., Mau, A., Nomikou, E. & Seakins, A. (2016). *Science capital made clear, an enterprising science publication.* https://www.stem.org.uk/sites/default/files/pages/downloads/Science-Capital-Made-Clear.pdf

Archer, L., Osborne, J., DeWitt, J., Dillon, J., Wong, B., & Willis, B. (2013). *Young people's science and career aspirations, age 10–14.* Aspires.

Berridge, G. (2011). *Event design and experience* (2nd ed.). Butterworth-Heinemann.

Brown, A., Donne, K., Fallon, P., & Sharpley, R. (2019). From headliners to hangovers: Digital communication in the British rock music festival experience. *Tourist Studies*, *20*(1), 75–95. https://doi.org/10.1177/1468797619885954

Bultitude, K. (2014). Science festivals: Do they succeed in reaching beyond the 'already engaged'? *Journal of Science Communication*, *13*(4), 1–3. https://doi.org/10.22323/2.13040301

Cabanas, E. (2020). Experiencing designs and designing experiences: Emotions and theme parks from a symbolic interactionist perspective. *Journal of Destination Marketing & Management*, *16*, 100330. https://doi.org/10.1016/j.jdmm.2018.12.004

Canovan, C. (2019). 'Going to these events truly opens your eyes'. Perceptions of science and science careers following a family visit to a science festival. *Journal of Science Communication*, *18*(02), 1–18. https://doi.org/10.22323/2.18020201

Canovan, C. (2020a). More than a grand day out? Learning on school trips to science festivals from the perspectives of teachers, pupils and organisers. *International Journal of Science Education, Part B: Communication and Public Engagement*, *10*(1). https://doi.org/10.1080/21548455.2019.1680904

Canovan, C. (2020b). Sharing the pi: Are incentives an effective method of attracting a more diverse science festival audience? *International Journal of Science Education, Part B: Communication and Public Engagement*, *10*(3). https://doi.org/10.1080/21548455.2020.1753126

Canovan, C., & Walsh, R. (2020). A space to study: Expectations and aspirations toward science among a low-participation cohort. *Journal of Science Communication*, *17*(2). https://doi.org/10.22323/2.19060204

Carneiroa, M. J., Eusébioa, C., Caldeirab, A., & Santos, A. C. (2019). The influence of eventscape on emotions, satisfaction and loyalty: The case of re-enactment events. *International Journal of Hospitality Management*, *82*, 112–124. https://doi.org/10.1016/j.ijhm.2019.03.025

Colman, E., Wanat, M., Goossens, H., Tonkin-Crine, S., & Anthierens, S. (2021). Following the science? Views from scientists on government advisory boards during the COVID-19 pandemic: A qualitative interview study in five European countries. *BMJ global health*, *6*(9), e006928.

DeWitt, J., Archer, L., & Mau, A. (2016). Dimensions of science capital: Exploring its potential for understanding students' science participation. *International Journal of Science Education*, *38*(16), 2431–2449. https://doi.org/10.1080/09500693.2016.1248520

Erduran, S., Ardac, D., & Yakmaci-Guzel, B. (2006). Learning to teach argumentation: Case studies of pre-service secondary science teachers. *Eurasia Journal of Mathematics Science and Technology Education, 2*(2), 1–14. https://doi.org/10.12973/ejmste/75442

Eshach, H. (2007). Bridging in-school and out-of-school learning: Formal, nonformal, and informal education. *Journal of Science Education and Technology, 16*(2), 171–190. https://doi.org/10.1007/s10956-006-9027-1

Fallon, P., & Robinson, P. (2017). 'Lest we forget': A veteran and son share a warfare tourism heritage experience. *Journal of Heritage Tourism, 12*(1), 21–35. https://doi.org/10.1080/1743873X.2016.1201087

Ferdinand, N., & Williams, N. (2018). The making of the London Notting Hill Carnival festivalscape: Politics and power and the Notting Hill Carnival. *Tourism Management Perspectives, 27*, 33–46. https://doi.org/10.1016/j.tmp.2018.04.004

Fogg-Rogers, L., Bay, J., Burgess, H., & Purdy, S. (2015). 'Knowledge is power': A mixed-methods study exploring adult audience preferences for engagement and learning formats over 3 years of a health science festival. *Science Communication, 37*(4), 419–451. https://doi.org/10.1177/1075547015585006

Gannon, M., Taheri, B., & Olya, H. (2019). Festival quality, self-connection, and bragging. *Annals of Tourism Research, 76*, 239–252. https://doi.org/10.1016/j.annals.2019.04.014

van Griethuijsen, R., van Eijck, M. W., Haste, H., den Brok, P., Skinner, N., Mansour, N., Gencer, A., & BouJaoude, S. (2015). Global patterns in students' views of science and interest in science. *Research in Science Education, 45*, 581–603. https://doi.org/10.1007/s11165-014-9438-6

Halat, E. (2007). Reform-based curriculum & acquisition of the levels. *Eurasia Journal of Mathematics Science and Technology Education, 3*(1), 41–49. https://doi.org/10.12973/ejmste/75373

Home Office. (2021). *Critical incident management, Version 13.0.* Home Office. https://assets.publishing.service.gov.uk/government/uploads/system/uploads/attachment_data/file/1013914/national-critical-incident-management-guidance-v13.0-ext.pdf

Jensen, E., & Buckley, N. (2014). Why people attend science festivals: Interests, motivations and self-reported benefits of public engagement with research. *Public Understanding of Science, 23*(5), 557–573. https://doi.org/10.1177/0963662512458624

Jepson, A., & Clarke, A. (Eds.) (2014). *Exploring community festivals and events.* Routledge.

Kennedy, E. B., Jensen, E. A., & Verbeke, M. (2017). Preaching to the scientifically converted: Evaluating inclusivity in science festival audiences. *International Journal of Science Education, Part B, 8*(1), 14–21. https://doi.org/10.1080/21548455.2017.1371356

Kerr, G. (2020). *Four shades of science festival: A qualitative study exploring the business and management dimensions of science festivals in the United Kingdom* [Doctoral dissertation].

Manning, C., Lin, K., & Goodman, I. F. (2013). *The Science Festival Alliance: Creating a sustainable national network of science festivals—Final summative evaluation report.* March. https://www.informalscience.org/sites/default/files/2013-08-09_SFA_2010-2012_Final_Evaluation_Report.pdf

McLoughlin, A. (2015). The future of event design and experience. In I. Yeoman, M. Robertson, U. McMahone-Beattie, E. Backer, & K. A. Smith (Eds.), *The future of events and festivals* (pp. 236–250). Routledge.

National Science Foundation. (2016). *Collaborative research: Broad implementation of Science Festival Alliance.* https://www.nsf.gov/awardsearch/showAward?AWD_ID=1223256

Osborne, J., Simon, S., & Collins, S. (2003). Attitudes towards science: A review of the literature and its implications. *International Journal of Science Education, 25*(9), 1049–1079. https://doi.org/10.1080/0950069032000032199

Packer, J., & Ballantyne, J. (2011). The impact of music festival attendance on young people's psychological and social well-being. *Psychology of Music, 39*(2), 164–181. https://doi.org/10.1177/0305735610372611

Peterman, K., Verbeke, M., & Nielsen, K. (2020). Looking back to think ahead: Reflections on science festival evaluation and research. *Visitor Studies, 23*(2), 205–217. https://doi.org/10.1080/10645578.2020.1773709

Quinn, B. (2006). Problematising 'festival tourism': Arts festivals and sustainable development in Ireland. *Journal of Sustainable Tourism, 14*(3), 288–306. https://doi.org/10.1080/09669580608669060

Rose, K. M., Korzekwa, K., Brossard, D., Scheufele, D. A., & Heisler, L. (2017). Engaging the public at a science festival. *Science Communication, 39*(2), 250–277. https://doi.org/10.1177/1075547017697981

Tum, J., Norton, P., & Wright, J. N. (2006). *Management of event operations*. Elsevier Butterworth-Heinemann.

University of Central Lancashire. (2021). *Lancashire Science Festival returns for a one-day spectacular*. https://www.uclan.ac.uk/news/lancashire-science-festival-returns-for-a-one-day-spectacular

Zhang, C. X., Hoc Nang Fong, L., Li, S., & Ly, T. P. (2019). National identity and cultural festivals in postcolonial destinations. *Tourism Management, 73*, 94–104. https://doi.org/10.1016/j.tourman.2019.01.013

3 Witch City's 'Haunted Happenings'

Managing Authenticity and Edutainment in Salem, Massachusetts

Hannah Stewart, Louise Todd and Jane Ali-Knight

Introduction

Festivals play a pivotal role in the development, marketing plans, and cultural outputs of many destinations. Festival tourism enables the promotion and animation of destinations and has become a significant area of research pertaining to urban redevelopment, education, and social change. Many festivals can be found within the context of dark tourism – tourism activities associated with the seemingly macabre (Stone, 2006). As such, dark festivals often favour the visual and experiential over historicity, thereby providing opportunities for diverse audiences to indulge in the macabre through emotionally evocative and edutaining interpretations of heritage and [sub]culture. For example, Dia de los Muertos is a religious festival that reunites the living with the dead, offering festival goers a nuanced look at the cultural exchange of mourning for celebration. Dark festivals remain largely under-researched within the context of their use of edutainment and their ability to engage audiences with culture and heritage. This chapter therefore explores the use of edutainment in dark festivals and does so through a case study of the Haunted Happenings festival in Salem, Massachusetts, USA, which uses edutainment to mix commercial Halloween experiences with the traditional elements of a harvest festival. Within this, this chapter addresses how the festival incorporates Salem's visitor attractions within the overall festival experience to better connect audiences to the local history and heritage.

Dark festivals and cultural heritage

The experience of cultural heritage within dark festivals is co-constructed through a symbiosis of contemporary demands and historic offerings – what people look for in the past shapes what we – as a society – know and what we remember of it (or have forgotten) (Simone-Charteris et al., 2016). However, whether a historical event is selected (and memorialised) for interpretation is reflective of the scale and uniqueness of the story (White & Frew, 2016). When these stories circumnavigate darker histories and death, the typology offered by Stewart and Stone (2022) may be referenced (Figure 3.1). Inferred by their

DOI: 10.4324/9781003305415-4

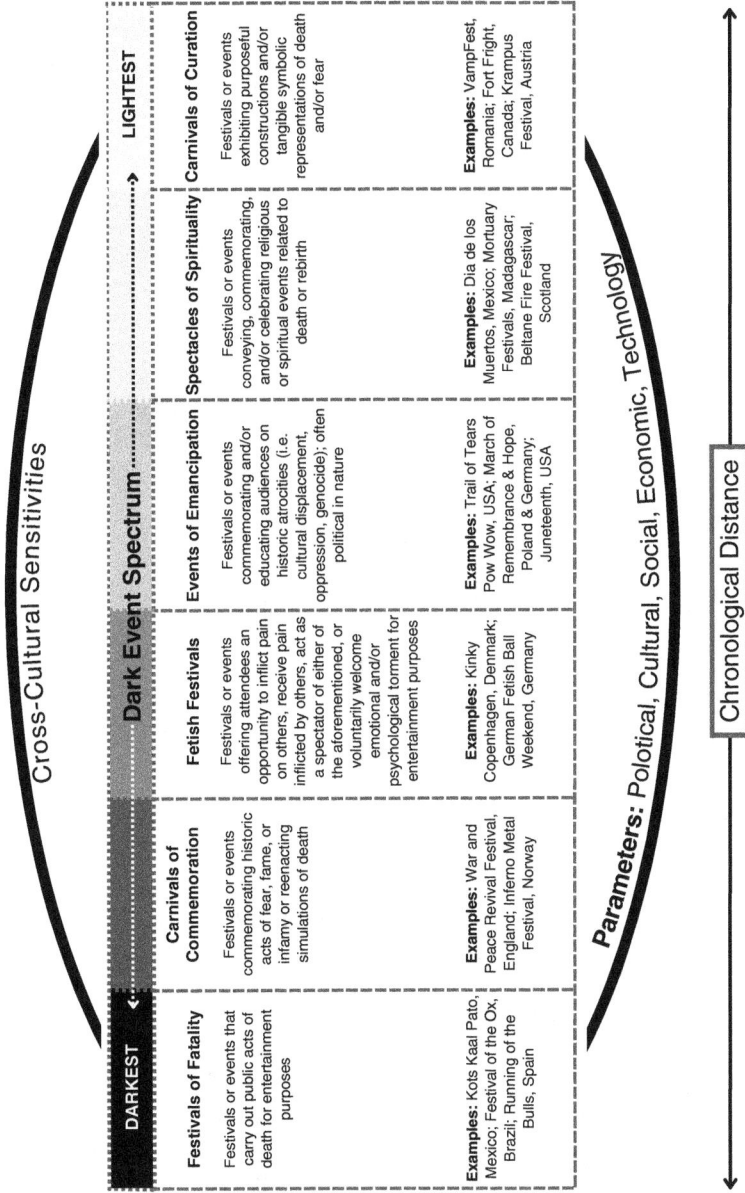

Figure 3.1 Parameters for Dark Event Tourism (DET): A typology illustrating supply features of DET within a darkest-lightest framework (after Stone, 2006).

Source: (Stewart & Stone, 2022).

associations with fatality or the macabre and socio-cultural, political, economic, and technological parameters, the Dark Event Tourism model illustrated in Figure 3.1 catalogues qualifying events within a darkest to lightest
framework (after Stone, 2006). Notwithstanding the semantics of dark versus
light dichotomies, aside, it offers a nuanced understanding of dark festival
offerings, perceptions, and characteristics; the way each story is [re]presented –
and subsequently interpreted – is integral to the overall visitor experience.
Appearing at the lightest end of the spectrum, the model posits Salem's annual
Haunted Happenings as a 'carnival of curation', and is defined as "festivals or
events exhibiting purposeful constructions and/or tangible symbolic representations of death and/or fear" (Stewart & Stone, 2022).

Dark festivals allow a finite window of opportunity to provide engaging and
edutaining experiences for festival goers. Whilst the supply–demand–manage
relationship is maintained, the symbiosis of interpretation efficacy and execution must be precise to maximise the potential to evoke appreciation and safe-
guard heritage assets within the overall festival experience (Beck & Cable,
2002; Wyatt, 2019) and ensure the long-term preservation of cultural heritage.

Edutainment and managing authenticity

Within dark festivals, edutainment is a catalyst for the interpretation of information, using entertainment, technology, and/or media to augment visitor
experiences. In practice, edutainment relies on interactive and immersive storytelling and theming, utilising stage design, performance theory, and execution
of Pine and Gilmore's (1998) Experience Economy to create immersive and/or
non-immersive experiences (Wyatt et al., 2021). In application, the Experience
Economy is guided by the 'Disney effect' and founded on the idea that a business stages its environment based on a clear, captivating storyline.

Seen often as a product of cultural, social, and political positioning, the
consumption of cultural heritage through dark festivals offers an opportunity
to satiate consumers' desire for authenticity and education within immersive
experiences (Teare & Lennon, 2017). Authenticity has the aptitude to inform
both cultural heritage and dark tourism destinations. Within the context of
tourism, authenticity is a feeling and/or emotion to be experienced in relation
to a place. Chhabra (2022) highlights the existence of four types of authenticity: essentialist (true to the origin), constructivist (commodified for monetary
gain), existentialist (optimal and euphoric), and negotiated (p. 22).

Authenticity can be influenced by external factors, including historical texts
(fictional and non-fictional) and pop-culture representations (Saxton, 2020).
The onset of image branding, image identity, and place endorsement has
encouraged the theatricality of almost any inanimate thing, including authenticity (MacCannell, 1973). Often contested is the relationship between authenticity and themed spaces (Milman, 2013). Theming is an exercise of staging that
combines both structure and organisation (Scheurer, 2004), resultant in delivering a unified experiencescape. The incorporation of thematic design elements

fosters cohesion throughout an overall event experience (Getz, 2010), proposing that an event's theme should be seamlessly reflected within all aspects of the event and its design (Allen et al., 2011).

Halloween, Inc.

Whilst some dark festivals are themed around commemoration (e.g. War and Peace Revival Festival, UK, Inferno metal festival, Norway), others centre around the fantastical oddities of cultural traditions (e.g. Krampus Festival, Austria, VampFest, Romania). An expansion of the latter, this chapter explores how Salem's Haunted Happenings uses the theme of Halloween to edutain October audiences on the city's storied past.

As a nation of immigrants, America's identity is entrenched in the adoption of its melting pot mentality (Smith, 2012). Perpetuated through time, cultural traditions have evolved into the American holidays celebrated today. Halloween is one such holiday, worth 10 billion USD to the American economy (Tighe, 2021). Halloween has roots in Celtic cultural tradition. According to the Pagan calendar, 31 October signifies the end of the harvest season and beginning of winter, which was marked by Samhain – a Druid festival of the natural and supernatural. As such, ethereal beings such as ghosts, fairies, witches, and demons became associated with the day. Observance of this evolved, and Halloween was brought to America in the 19th century by Irish and Scottish immigrants, many of whom continued their harvest traditions.

Halloween has since established itself as a major event and all-American family holiday, adopting the act of carving pumpkins, donning costumes, and trick-or-treating in anticipation of the one night each year the dead may return to earth. Touted as Halloween capital of the world (Haunted Happenings, 2022), Salem's annual festival has become a mecca for anyone interested in everything from history to Halloween and everything in between.

Witch City, USA

Yet, both Destination Salem and local community factions wish to loosen ties with the city's overpowering witch history (Chevedden, 1997). This was reflected in a 2010 advert campaign – *There's More to Salem*. This type of interplay confirms the dilemma within undesirable heritage (MacDonald, 2006). Salem, however, is not the only location familiar with the 'identity versus economy dilemma' – 'UFO Tourism' is a major economic contributor to Roswell, New Mexico, USA, in the same way that 'Nuclear Tourism' is to Chernobyl, Ukraine. This proves challenging for the countries and/or cities involved, as it sensationalises aspects of history that may be unfavourable, forcing a nuanced and at times singular interpretation of a place or time in history.

Founded in 1626, Salem, Massachusetts, has become the epicentre of witchcraft and Halloween due to a collection of historical events that occurred independently. Rooted in Salem, the Witch Trials do not hold monopoly over its occultist standing. A version of these trials was revived with Arthur Miller's 1953 play *The Crucible*, a satire on the danger of the McCarthy-era witch hunt for communists within the American government and again during the unveiling of the Salem Witch Trials Memorial at the Tercentenary event in 1992. Trials aside, the 1900 publication of L. Frank Baum's *The Wizard of Oz* debuted two of pop culture's most iconic archetypes – the Wicked Witch of the West and Glinda the Good Witch. Historically a symbol entrenched in Satanic worship, the book's success saw the witch transmogrify into a palatable caricature of her former self. In the 1970s, witches further evolved into a household entity with the airing of the Salem Saga episodes of *Bewitched,* which enticed viewers to accept witchcraft as a wholesome social movement, thus removing its exclusivity to Satan's disciples. Today, however, the stereotype of a warty-nosed hag, donning a conical hat, and riding a broomstick with her black cat has solidified the symbolic theatricality linking witches (and subsequently Salem) to Halloween (Amin, 2019).

Haunted happenings – A case study

Salem offers a plethora of year-round interactive museums, tours, attractions, and historic homes for visitors to learn more about the city's maritime past and witch history and how they continue to shape Salem's cultural and social landscapes. Yet, America's fixation with Halloween and Salem's centuries-old connection to the supernatural has perpetuated the city's synonymity with America's favourite holiday; as global intrigue of the witch as a pop-culture icon increased, so did tourists to the city. Unfortunately, 'Witch City' had a name but little entertainment to offer its October visitors. The first Haunted Happenings festival emerged during the 1982 Halloween weekend and aimed to enhance the city's tourism offerings, providing visitors with family-friendly entertainment. Regarded as one of the world's top Halloween festivals (National Geographic, 2018), Salem transforms into New England's own Halloween Town every October. While some aspects of the festival offer a denatured sanctification of the witch trials, others offer a 'Disneyfied' interpretation of its history (Nugua, 2006) via trial re-enactments and caricatures offering paid photo opportunities on the main street. Annually, festival goers account for more than half the city's yearly tourist traffic, which is reflective of the seductive allure of fantasy, escapism, and edutainment (Pine & Gilmore, 1998).

While some dark festivals favour purpose-built venues, others rely on existing structures and product offerings as the foundation of the experience. With tourism as Salem's primary industry, Haunted Happenings augments its year-round touristic offerings with accentuated opportunities each October (see Table 3.1). These year-round visitor attractions and the various activities that take place within them are the major delivery arm of the festival, while the namesake of the festival acts as a catalyst that attracts additional visitors for the city-wide Halloween celebrations. Festival-specific attractions include the Salem Chamber of Commerce Haunted Happenings Grand Parade, which

Table 3.1 Festival attractions and festival events, *Haunted Happenings* 2022

		Examples
Haunted Happenings: Official Event Guide Categories, October 2022	**Special Events**	Free Family Movie Nights; Salem Harvest Fest; Kid's Fun Fest; Haunted Speakeasy; The Salem Haunted Magic Show; *Haunted Happenings* Marketplace; Guilty! A Salem Witch Trial; The Official Salem Witches' Halloween Ball; Halloween Finale Fireworks
	Museums & Attractions	The Salem Witch Museum; Gallows Hill Museum/ Theatre; House of the Seven Gables; New England Pirate Museum; Halloween Museum; Peabody Essex Museum; Salem Heritage Trail; The Satanic Temple
	Haunted Houses	Frankenstein's Castle; Chambers of Terror; The Lost Museum at Gallows Hill; Witch Mansion
	Harbour Tours	Eerie Stories on Salem Sound; *Haunted Happenings* Cruise; Five Lighthouses & Foliage Cruise
	Trolley Tours	Tales & Tombstones; Ghosts & Legends Trolley at Gallows Hill
	Walking Tours	Spellbound Tours; AM Coffee Walks & Salem Food Tours; Bewitched After Dark Tours; Tragedy & Triumph: Tales of Blood and Treasure in Salem
	Psychics, Psychic Fairs & Seances	Omen: Psychic Parlour & Witchcraft Emporium; The Salem Psychic Fair and Witches' Market; Maison Vampyre; Salem Séance
	Regional Activities	The Rebecca Nurse Homestead; Hammond Castle Museum; North of Boston Convention & Visitors Bureau
	Restaurants & Bars	Flying Saucer Pizza Company; Reds Sandwich Shop; OPUS; Salem's Retreat; Witch City Hibachi; Turner's Seafood at Lyceum Hall
	Sweets & Treats	Odd Metre Coffee Co.; Harbour Sweets; Coffee Time Bake Shop; Kakawa Chocolate House
	Shopping	Artists' Row; Bewitched in Salem; Coven Dark Boutique; Die With Your Boots On; Circle of Stitches; The Black Veil Shoppe of Drear & Wonder; HausWitch Home + Healing; Wynott's Wands
	Witch & Occult Shops	Artemis Botanicals; The Cauldron Black; Hex: Old World Witchery; Crow Haven Corner; The Witchery – Broom Making, Bookbinding, Shows
	Accommodation	Clipper Ship Inn; Hawthorne Hotel; Harbour Light Inn; Holiday Inn & Suites Boston Peabody

[*Source*: Haunted Happenings, 2022].

marks its official opening while other seasonal favourites include the Howl-o-ween Pet Parade, Mayor's Night Out, and Hawthorne Hotel's Halloween Ball (see https://www.hauntedhappenings.org/2022Guide/for a comprehensive list of 2022 activities).

Methods

Using a case study approach, this research explores the use of edutainment on offer throughout Haunted Happenings and how the marriage of augmented annual attractions turned annual festival attractions and seasonal stand-alone events connect audiences to Salem's infamous history. Part of the author's post-graduate study (2019), the original research investigated dark tourism, festival and event tourism, and place identity to better understand the role of Haunted Happenings in forming Salem's identity. This included exploratory research across two October festival attractions through purposive sampling – The Salem Witch Museum (SWM) and SpellBound Tours (SBT). Although both are annual-attractions-turned-annual-festival-attractions, these are two of the most popular festival events with festival goers across all October offerings. They were selected because of the researcher's ability to confirm interview appointments with their owners/operators; echoed by all interview participants was how busy October was for the tourism sector. Special events were eliminated as case study contenders due to budgetary restraints and ticket availability. The research sample highlights the exceptionality of the 2019 Haunted Happenings festival in delivering an edutaining interpretation of Salem's history, associations with the occult, and ties to Halloween within a controlled environment.

Data collection involved a combination of 13 semi-structured interviews (see Table 3.2) and netnographic research. This permitted the study to be conducted both actively (in person) and passively (with digital communities). TripAdvisor and Facebook business account pages were used where a total of 7,370 comments were examined. All social media handles and usernames were changed to ensure anonymity and confidentiality and to prevent any ethical issues. To discourage the possibility of deception, the researcher involved participants at significant landmarks throughout the write-up process to ensure individual perspectives were authentically and acceptably portrayed. Prior to data collection, the researcher assessed and/or neutralised any potential psychological, social, and/or physical risks to participants through festival attraction-specific risk assessments, and respondent information was only collected once each informant completed a consent form. Interview questions were formulated based on the literature that discussed the role of cultural heritage, festivals, dark tourism, and managing authenticity in creating edutaining outputs during the Haunted Happenings festival.

Following consent, interviews were conducted, lasting 60–70 minutes, digitally recorded, and subsequently transcribed. Transcriptions were manually coded through line-by-line review using elements of both thematic and content analysis. This was embedded in a grounded theory approach allowing for multiplicity in data sourcing with participant observations, interviews, and secondary research methods to complement one another (Goulding, 2002). The researcher chose a narrative approach to the subsequent analysis and discussion of the case studies to capture the atmosphere of mystery and storytelling used to edutain festival audiences throughout the Haunted Happenings festival.

Table 3.2 Interviewee profiles, semi-structured interviews

Interviewee	Gender	Age Range	Demographic	Industry	City of Residence (Massachusetts)	Festival Involvement (Years)
Interviewee 1	Female	25–34	Practising Witch	Tourism	Salem	11
Interviewee 2	Female	35–44	Advertising	Tourism/Education	Groveland	20+
Interviewee 3	Female	18–24	Institutional Educator	Tourism/Education	Beverly	5+
Interviewee 4	Female	18–24	Marketing Intern	Tourism/Education	Salem	5+
Interviewee 5	Male	35–44	Entrepreneur	Tourism	Salem	10+
Interviewee 6	Male	35–44	Entrepreneur	Tourism	Salem	11
Interviewee 7	Female	18–24	Public Relations	Tourism/Education	Salem	5+
Interviewee 8	Male	35–44	Entrepreneur	Tourism	Salem	17
Interviewee 9	Female	18–24	Resident	Hospitality	Salem	20+
Interviewee 10	Male	35–44	Practising Witch	Tourism	Salem	15+
Interviewee 11	Female	45–54	Entrepreneur	Tourism	Boston	10
Interviewee 12	Female	35–44	Business Development	City Official	Salem	20+
Interviewee 13	Male	55–64	Historian	Archives	Danvers	20+

Salem Witch Museum

The Salem Witch Museum has been welcoming visitors since 1972. Its original owners are responsible for the creation of the Haunted Happenings festival and are ranked as the number one festival attraction to visit during October (Destination Salem, 2022) because of its reputation for being one of few outputs to deliver a factual interpretation of historical events. While the physical exhibits themselves are not altered for October visitors, the manner in which festival goers are encouraged to interact with them, the SWM's approach to crowd management and gift shop offerings are. Festival goers must purchase timed entry tickets prior to visiting, as the event is likely to sell out. The museum offers two exhibitions in which the 1692 Trials are portrayed and analysed. The initial presentation theatrically illuminates staged tableaus in period costumes while a dramatic narration retells the tragic events leading to the deaths of 20 innocent people. Running at 30-minute intervals, festival goers are ushered into the middle of a dark auditorium with theatre-in-the-round-style seating. A red circle on the floor casts a Satanic glow in the centre of the room, revealing the victims' names in concentric circles. The story begins. Lucifer, himself, with a sepulchral tone of narration, expresses the Puritanical views and religious fear of the Devil in 17th-century New England. The scene changes to Betty Parris – Reverend Parris' nine-year-old daughter, and her friend Abigail Williams sitting in front of the fire with Betty's nurse, Tituba. A booming recitation tells of Tituba sharing tales of the Devil, satanic worship, and practising voodoo as a child in South America. Soon after, Betty falls ill; other school-aged girls in Salem become likewise 'afflicted' – it *must* be witchcraft. The witch hunt begins. Following, a dozen or so dioramas are sequentially lit and narrated, immersing guests in the world of 1692 Salem. At the conclusion of narration, half of the guests file into the second exhibition while the other half enter the gift shop to await entry to the former. October gift shop offerings differ from the rest of the year placing emphasis on Halloween merchandise, including Witch Kitsch, literature (both historical and fantastical), and display cases showcasing historical 'Witch City' artefacts from the late 19th century to present.

The second festival exhibition, 'Witches: Evolving Perceptions', explores the evolution of the term 'witch', examines the role of European witchcraft trials on Western Society, and pop culture, and critically analyses the witch-hunting phenomenon. Here, people collectively engage with each presentation as a museum employee and tannoy recording guides guests through the exhibit. Tableaus depicting the roles and perceptions of witches throughout history are situated amidst the timeline. From local healer, to the portrayal of *The Wizard of Oz* witches, to Salem's modern witch residents, it provides autonomy to a marginalised community and highlights the injustice of the witch era.

The official Haunted Happenings festival programme advertises the SWM as Salem's most visited museum, proclaiming, "there's no better place to be on

Halloween!" (Haunted Happenings, 2022). In addition, during the festival, the Museum facilitates additional discussions on the pop-culture evolution of the witch's broom and conical hats to extend their audience reach and further educate visitors on Salem's witch history. On this topic, one interviewee explained:

> We are the constant driver of educational activities during Haunted Happenings because a lot of people come here for Halloween who wouldn't have come otherwise. Then, they accidentally have this educational experience which is a very interesting piece of the Haunted Happenings phenomena. People come [to the museum] and are surprised by the experience they have because it's unique and entertaining. During October, we showcase Salem's history in a manner that celebrates witchcraft as a cultural site where pointed hats, broomsticks and even the green face aren't part of a tragedy, but instead a fascinating part of our cultural evolution.
>
> (SWM, Interviewee 3)

The Salem Witch Museum is, thus, the purveyor of factual Salem witch history (and beyond) edutainment for festival goers during Haunted Happenings.

Salem's history unintentionally lends itself to being a mecca for all things 'spooky', particularly during Halloween – "Oftentimes, later in the evening, visitors will think [the museum] is a haunted house because of its Gothic Revivalist structure and they'll queue up for hours just to get in without knowing what they're waiting for" (SWM, Interviewee 10). Even the greatest nonbelievers will likely learn something because of the environment curated across Salem and its October attractions. This evidentiates Haunted Happenings' ability to aptly interpret meaningful historical events and provide memorable, engaging, and edutaining experiences for visitors at the museum.

SpellBound Tours

Proclaimed as Salem's original and best-haunted tour, SpellBound Voodoo, Vampires and Ghosts Tour is ranked #5 out of 72 tours and activities offered across the Haunted Happenings festival programme, and, as of October 2022, had 1,834 five-star reviews on TripAdvisor. It boasts an opportunity for festival goers to "learn what really happened" during the infamous Salem Witch Trials, where all tours are led by trained paranormal investigators. Lasting approximately 90 minutes, festival goers visit 12 sites pertinent to Salem's witch hysteria and the lesser-known Voodoo roots of Tituba, New England's vampiric past, and rife with documented paranormal activity.

The guide sets the scene before the tour begins. Clothed in a black suit, he sports horn-rimmed spectacles, lamb-chop sideburns, and a handlebar moustache; he dons a top hat with a shrunken head made from goat skin, human bones attached to the front. Hands folded atop his skull-adorned swagger stick, he begins:

> On this tour, people have been scared, people have fainted, and people have had ... experiences – one's they internalise and that follow them home. I don't know what you believe in, you may not even believe a word I say, but I encourage you to look it up and fact-check all that I've shared with you when you get home.
>
> (SBT, Interviewee 6)

The guide maintains his character as a charismatic paranormal historian throughout the tour and focuses his stories on historical accuracy; he embellishes with frightful fanfare and encourages guests to photograph the space in anticipation of capturing a paranormal presence – he interweaves storytelling with personal anecdotes to edutain festival goers. In the courtyard of Town Hall, he discusses Salem's ties to Vampirism and Voodoo, exposing Salem's lesser-known community of self-identified vampires who yearn for human blood and are accepting donor applications. At the tour's conclusion, guests examine and share their touristic photos, as they may have captured a ghostly orb or auratic figure. A fellow tour goer examines a photo taken outside of St. Peter's Episcopal Church, one of the stops where the gravestones of the plots had been separated from their bodies and relocated to the churchyard, alone. Aghast, the guide traces his finger over what he believes is the presence of a spectral entity. Was this truly proof of a paranormal presence, or simply the product of dimly lit streets and optics?

SpellBound Tours' ability to provide an evocative and transcendent experience for ticket holders is not only testament to the tour's 96% recommendation rate on TripAdvisor but emphasises the curated exceptionality of Haunted Happenings events. Yet, the Director of Operations acknowledges the challenges associated with his tour's popularity throughout the festival:

> Unfortunately, from my perspective, [the tour] becomes less about edutainment and becomes more about crowd control. Sometimes the route changes out of necessity because there are so many [other walking tours], and while I try to dress flashier – I put LEDs on my hat and carry a big lit-up staff for visibility – I often end up bundled up because it gets so cold!
>
> (SBT, Interviewee 6)

Destination Salem's Director echoes this point, offering that the event management strategy for Haunted Happenings is broken down into 31 days and one night – "as soon as 4pm hits on the 31st, [the festival] becomes more of a safety effort and less of a marketing initiative" (Interviewee 12). Of 1,020 TripAdvisor reviews, 28 are exclusive to October enthusiasts. Visitors describe their experience as "fun and informative" (Joe, TripAdvisor, October 2019), with "lots of information and fun content on an engaging and well-run tour" (Kate, TripAdvisor, October 2019) – the type of experience where you, "come for the stories, stay for the entertainment and leave having learned a ton about the history of Salem as a whole" (Mike, TripAdvisor, October 2019). SpellBound Tours effectively marries education with entertainment and finds a way to engage October festival goers with the regaling of legends and folklore within a historical context pertinent to Salem's heritage and culture.

Discussion

By the 20th century, witchcraft had evolved into a non-threatening, easily marketable destination brand for Salem. Presently, Destination Salem markets the city around three seasons: January through September, October, and November through January (Interviewee 7, 2019). These campaigns aim to generate a brand identity unique to Salem, distinguishing it from other destinations and highlighting its unique cultural offerings, seasonally. Destination Salem's use of witch iconography conjures recognisable thoughts and emotions culturally significant to Salem's portrayed identity and social milieu, subsequently giving depth to the semiotic landscape of the city's cultural heritage and visitor interpretation of its history. Salem's storyline is married to its witch history and embodiment of Halloween, providing the city with a unique advantage over other destinations in a highly competitive market.

The festivalisation of Salem's supernatural narrative is commodified through a culmination of event-specific offerings under the Haunted Happenings umbrella – the festival is testament to a cultural desire to engage with the past in creative ways. Some business owners recognise the opportunity for accentuated dividends by satiating the 'witchcraft Disneyland' expectations of dark festival tourists. Conversely, some historians and the represented witch community aim to educate and "...decry pervasive witchcraft imagery as indecorous" (Olbrys-Gencarella, 2007, p. 271). Haunted Happenings provides opportunities for 'mediated resistance', a strategy involving the tolerance of the anti/pro-tourism dichotomy, with the greater goal of protecting Salem's cultural integrity, identity, and authenticity (Joseph & Kavoori, 2001).

On the basis of the two annual-attractions-turned-annual-festival-attractions explored throughout the case study, the successful thematic integration of Halloween and witch historicity throughout Haunted Happenings embodies this performative form of 'staging' (Gran, 2010), offering festival goers meaningful and transcendent experiences. In the context of edutainment, staging is a "process that unifies, education, entertainment and technology through strategic organisation and structure" (Wyatt et al., 2021, p. 435). Although the feat of measuring value and authenticity is impossible, edutainment opportunities throughout the festival are authenticated through locals and visitors' "emotional, affective, sensuous, and relatedness" to product offerings (Knudsen & Waade, 2010, p. 13).

Conclusion

As a destination, Salem epitomises the allure of experience design, reflected through its dark festival attractions and their ability to edutain and connect festival goers to Salem's history and culture throughout Haunted Happenings. From a demand perspective, Salem's tourism sector plays to its strengths, using the 'Disneyfied' expectation of the city as a catalyst to impart the *real* history of Salem, its affinity with witchcraft, and ties to Halloween through engaging and immersive interpretations and events.

Despite the elevated tourism infrastructure, commercialisation, and temporal distance of the 1692 Trials, there is recognition among the sample festival attractions of the importance of providing an immersive, fact-based, and edutaining experience. Haunted Happenings uses a range of festival experiences to expand the festival audience and reach of the festival and further convey its education messaging. The festival aims to engage with the past in a real and authentic way through the provision of immersive and memorable experiences for festival goers. To perpetuate this, the festival engages with the Salem Witch Museum, which uses evocative and immersive messaging, staged tableaus, and storytelling to encourage visitors to critically consider the cultural and social ramifications of unjust political and social actions in hopes of mitigating the risk of future witch hunt outbreaks, and SpellBound Tours, which activates visitation to sites of tragedy or the macabre and enhances the impact of each site through the sharing of historical facts, interwoven with injections of local legends, folklore, and first-hand encounters with the paranormal. Both use contemporary references to narrow the spatial distance and liminality of Salem's dark history, making it more relatable and relevant for festival attendees. This exploration evidentiates the importance of education and entertainment within dark festivals and offers a nuanced look at how festival creators actualise edutaining experiences for festival consumption.

Future research into dark festivals would benefit from the following explorations: (1) Which aspects of tragic history are most suitable for interpretation and staging?; (2) What is the most effective and respectful way to interpret/commemorate/animate the past?; (3) Who has jurisdiction over which events are selected and stories told?; (4) How is historical authenticity managed? Such considerations would offer a deeper understanding of the role of dark festivals in animating spaces through culturally authentic storytelling and interpretation of seemingly macabre histories.

Declaration

The contents of this chapter draw on a larger MSc research project conducted between May and August 2019 in Salem, Massachusetts, as part of the requirements for successful completion of the International Festival and Event Management programme at Edinburgh Napier University (UK).

References

Allen, J., O'Toole, W., Harris, R., & McDonnell, I. (2011). *Festival & special event management* (5th ed.). John Wiley & Sons Australia, Ltd.

Amin, A. (2019). Understanding the changing concepts of Halloween in America. *Digital Press Social Sciences and Humanities*, *2*, 1–6. https://doi.org/10.29037/digitalpress. 42252

Beck, L., & Cable, T. (2002). *Interpretation for the 21st century: Fifteen guiding principles for interpreting nature and culture* (2nd ed.). Sagamore Publishing.

Chhabra, D. (2022). *Resilience, authenticity, and digital heritage tourism*. Routledge.

Chevedden, P. (1997). Ushering in the millennium, or how an American city reversed the past and single-handedly inaugurated the end-time. *Prospects*, *22*, 35–67. https://doi.org/10.1017/S0361233300000041

Destination Salem. (2022). *Salem Witch Museum–Salem, MA*. https://www.salem.org/listing/salem-witch-museum/

Getz, D. (2010). *Event studies: Theory, research and policy for planned events* (2nd ed.). Butterworth-Heinemann.

Goulding, C. (2002). *Grounded theory a practical guide for management, business and market researchers*. Sage.

Gran, A. (2010). Staging places as brands: Visiting illusions, images and imaginations. In B. Knudsen, & A. Waade (Eds.), *Re-investing authenticity* (1st ed., pp. 22–37). Channel View Publications.

Haunted Happenings. (2022). *About Salem Haunted Happenings*. https://www.hauntedhappenings.org/about/

Joseph, C. A., & Kavoori, A. P. (2001). Mediated resistance: Tourism and the host community. *Annals of Tourism Research*, *28*(4), 998–1009. https://doi.org/10.1016/S0160-7383(01)00005-6

Knudsen, B., & Waade, A. (2010). *Re-investing authenticity: Tourism, place and emotions*. Channel View Publications.

MacCannell, D. (1973). Staged authenticity: Arrangements of social space in tourist Settings. *American Journal of Sociology*, *79*(3), 589–603. https://www.jstor.org/stable/2776259

MacDonald, S. (2006). Undesirable heritage: Fascist material culture and historical consciousness in Nuremberg. *International Journal of Heritage Studies*, *12*(1), 9–28. https://doi.org/10.1080/13527250500384464

Milman, A. (2013). Guests' perception of staged authenticity in a theme park: An example from Disney's Epcot's World Showcase. *Tourism Review*, *68*(4), 71–89. https://doi.org/10.1108/TR-09-2013-0052

National Geographic. (2018). *10 great Halloween celebrations in the United States*. https://www.nationalgeographic.com/travel/article/10-great-halloween-celebrations-in-the-u-s.

Nugua, A. (2006). Witch city and mnemonic tourism. *Journeys*, *7*(2), 55. https://doi.org/10.3167/jys.2006.070204

Olbrys-Gencarella, S. (2007). Touring history: Guidebooks and the commodification of the Salem Witch Trials. *Journal of American Culture*, *30*(3), 271–284. https://doi.org/10.1111/j.1542-734X.2007.00556.x

Pine, B., & Gilmore, J. (1998). Welcome to the experience economy. *Harvard Business Review*, *76*(4), 97–105. https://doi.org/10.4337/9781781004227.00007

Saxton, L. (2020). A true story: Defining accuracy and authenticity in historical fiction. *Rethinking History*, *24*(2), 127–144. https://doi.org/10.1080/13642529.2020.1727189

Scheurer, R. (2004). Theme park tourist destinations: Creating an experience setting in traditional tourist destinations with staging strategies of theme parks. In K. Weiermair & C. Mathies (Eds.), *The tourism and leisure industry: Shaping the future* (pp. 227–236). Research Institute for Leisure and Tourism, University of Berne.

Simone-Charteris, M., Boyd, S., & Burns, A. (2016). The contribution of dark tourism to place identity in Northern Ireland. In L. White & E. Frew (Eds.), *Dark tourism and place identity: Managing and interpreting dark places* (pp. 60–78). Routledge.

Smith, D. M. (2012). The American melting pot: A national myth in public and popular discourse. *National Identities*, *14*(4), 387–402. https://doi.org/10.1080/14608944.2012.732054

Stewart, H., & Stone, P. R. (2022) Festivalisation and modelling the macabre. Proposing a typology for Dark Event Tourism (DET). Presentation at the *Dark tourism research symposium: Memory, pilgrimage and the digital realm*. Edinburgh Napier University, Edinburgh, 5 May.

Stone, P. (2006). A dark tourism spectrum: Towards a typology of death and macabre related tourist sites, attractions and exhibitions. *Tourism: An Interdisciplinary International Journal*, *54*(2), 145–160.

Teare, R., & Lennon, J. J. (2017). Editorial. *Worldwide Hospitality and Tourism Themes*, *9*(2), 130–131. https://doi.org/10.1108/WHATT-01-2017-0001

Tighe, D. (2021). *Annual Halloween expenditure in the United States 2005-2021*. https://www.statista.com/statistics/275726/annual-halloween-expenditure-in-the-united-states.

White, L., & Frew, E. (2016). *Dark tourism and place identity* (2nd ed.). Routledge.

Wyatt, B. (2019). Creating truth through edutainment: An exploration of influences on the design and management of interpretation within lighter dark tourism practice. Presentation at the *AHI annual conference*, Bedford.

Wyatt, B., Leask, A., & Barron, P. (2021). Designing dark tourism experiences: An exploration of edutainment interpretation at lighter dark visitor attractions. *Journal of Heritage Tourism*, *16*(4), 433–449. https://doi.org/10.1080/1743873X.2020.1858087

4 Edutainment in Disney's EPCOT International Festival Cycle

Carissa Baker

Introduction

EPCOT, or the Experimental Prototype Community of Tomorrow, opened in 1982 as the second theme park at Walt Disney World in Florida, USA. Originally conceived of by Walt Disney as a city of the future, EPCOT was altered conceptually after Walt's death to be a permanent world's fair, with one half of the park a tribute to multiple nations (World Showcase) and the other a celebration of technology and progress (Future World). Walt Disney World, the largest and most visited theme park complex in the world, consists of shopping, dining, lodging, two waterparks, and three other theme parks: Magic Kingdom (1971), Disney's Hollywood Studios (1989), and Disney's Animal Kingdom (1998). A more educational than fantastical theme park, EPCOT eventually became coded as an adult offering amongst this family line up.

In an attempt to change the image but maintain the "soul" of the park with opportunities for discovery and education (Byrd, 1996), EPCOT established four annual festivals: the International Flower & Garden Festival (1994), the International Food & Wine Festival (1996), the International Festival of the Holidays (1996), and the International Festival of the Arts (2017). These festivals immediately lent to core business needs such as revenue, attendance, repeat visitation, and park image – all made possible through an edutainment design. As one of the top ten most visited theme parks in the world, with between 7 and 12 million visitors per year (Rubin, 2020, 2022), this chapter considers EPCOT's four festivals. Specifically, it explores how these festivals have helped to make EPCOT a theme park dedicated to edutainment and exploration while simultaneously emphasising several sorts of visitor consumption and helping to boost revenue, play, cultural immersion, and learning.

Edutainment

Walt Disney coined the term "edutainment" in the 1940s, describing his educational films with entertainment elements, a form common to the studio through its productions on topics such as nature, war, transportation, and outer space (Van Riper, 2011). Wright (2006) argued that edutainment is a fundamental

DOI: 10.4324/9781003305415-5

principle at Disney, claiming "one of the central themes of Walt Disney's body of work is the notion that learning can be fun and entertaining" (p. 56). This is unsurprising considering the success of Disneyland in 1955 (which included edutainment offerings like exhibits and films), but some theme parks have taken on an edutainment function, with none more so than EPCOT. Other parks with explicitly educational content include Disney's Animal Kingdom, Knott's Berry Farm, and Dollywood in the United States, Futuroscope and Puy du Fou in France, Europa-Park in Germany, and the Songcheng parks in China.

Cultural attractions such as museums have shifted to more theatrical, thematic, and immersive exhibits over the years (Kirshenblatt-Gimblett, 1998), indicating the growing popularity of edutainment in tourism. Heritage sites have likewise used edutainment, for example, atmospheric immersion and theatrical performance in Colonial Williamsburg, as an approach to immerse visitors in the past and encourage active participation (Kerz, 2016). Even dark tourism sites have used edutainment techniques to engage visitors with strategies such as character reenactment, set design, technologies, self-guided tours, and thematic storytelling (Wyatt et al., 2021).

Festivals may also have an explicit educational role, as in the instance of garden festivals that inform the public about flora and desire to influence environmental design behaviour (Ibrahim et al., 2015). They often do so with entertainment flourishes, thus becoming edutainment spaces themselves. While the festivals at EPCOT are more commercialised than those that overtly focus on science or community building, even small-scale festivals can promote educational content, which can then influence behavioural intentions (Tzaczyski et al., 2022).

The Value of Festivals

Festivals possess both intrinsic and extrinsic values that impact societies and individuals, including economic, individual, and community; they also hold inherent benefits and challenges (Getz et al., 2018). There are many kinds of festivals, with the festivals at EPCOT matching four of the types: arts festival, garden festival, food festival, and holiday festival. There are some obvious benefits to festivals, including preserving cultural heritage and providing cultural value to communities (Ascione & Fink, 2021), building a destination brand (Lee & Arcodia, 2011), and economic values such as added jobs, benefits to businesses, impact on the hospitality and tourism industries, and stimulating local economies (Getz et al., 2018). Within every festival context, there are diverse stakeholder perspectives from guests and employees to governments and communities of interest (Getz et al., 2018).

Festival features may attract visitors beyond only the consumption of food or performances. One study (Leer & Juel-Jacobsen, 2022) found that visitors encounter many facets of festivals with intellectual, sensory, and social experiences. Festivals would present these aspects with educational activities (learning opportunities such as workshops, demonstrations, readings, or quests), senses within the space (environmental design, consumption of food and beverage), and a social character (engaging with friends, family, and strangers). Theme parks are inherently sensory, dimensional, and social; thus, festivals are a natural addition to the theme park context. If motivated guests perceive festivals as high quality and are satisfied with the service, they are more likely to display loyalty through revisit intention and recommendations to others (Amorin et al., 2019).

Choo et al. (2022) noted that food festivals have more active participation than several other touristic activities. They posited a model wherein this involvement can lead to satisfaction and then loyalty, with several other factors being important including the products, environment, and meaning making. Performing arts festivals also have positive impacts, including the co-creation of value through multiple stakeholders from organisations to sponsors and artists to visitors (Chang, 2020). A positive impact can also include behavioural change. A previous study (Organ et al., 2015) found that engagement at food festivals can change food buying behaviour in the future. This would be a good thing to study at EPCOT as well, as anecdotally guests have more curiosity and adventurousness in future culinary experiences.

Consumption

Ritzer (2010) considered well-designed leisure destinations like Disney theme parks to be "cathedrals of consumption", referring to the spaces that enchant visitors to make consumption pleasurable (p. 7). Disney theme parks have been argued to narrativise consumption (Bryman, 1995), including the consumption of corporate brands and technologies, something found at multiple theme parks through sponsored products and exhibits (Martin, 2019). At EPCOT, visitors can consume the past, culture, and themed environments themselves (Houston & Meamber, 2011). In World Showcase, it is a consumption of national cultures (Knight, 2014) or "transnational cooperation and consumption" (Kratz & Karp, 1993).

During the annual cycle of the four festivals, there is a great deal of consumption, whether it be food, merchandise, branding, culture, or the physical space. Additionally, a fan culture has grown up around festival items, with numerous blogs (e.g. *The Disney Food Blog* or *Vegan Disney World*) focusing on food and food practices at Disney. Some fans may consider it a form of "culinary capital" that impacts individual and group identities (Naccarato & LeBesco, 2012). With Disney fans, in particular, social capital may be achieved by consuming special event foods, with the practice connected to prestige within the fan community (Williams, 2020). This is in line with Putnam's (1995) concept of social capital, a reputation afforded through networks, norms, and resources, with festivals having both social and financial components. It likewise aligns with Misener (2013), who determined that events facilitate social capital through enhancing individual skills, augmenting community organisations, and creating social links between people and organisations. Like with other festivals (Mair & Duffy, 2018), Disney's can assist with building both individual abilities and social links. In the Disney festivals context, increasing social capital has now extended beyond the actual food and beverage items to the associated merchandise. One of the most popular Disney-related stories on social media in 2022 was about the multiple-hours wait to purchase a popcorn bucket of the EPCOT character Figment during the International Festival of the Arts, with those unable to secure one paying up to hundreds of dollars on eBay (Martin, 2022). Souvenir purchases at Disney parks hold significance in several cultures (Wei, 2018), and the festival offerings have now contributed to this demand.

Finally, one discussion in tourism centres around guests desiring to consume experience, not just products. One study (Liang & Li, 2021) finds that hybrid consumption, with guests consuming everything from food to attractions, is an essential part of the theme park experience. Theme parks function within all the experience realms (entertainment, education, aesthetic, escapist), making them complex spaces for consumption (Lee et al., 2020). Festival goers likewise connect with experiential consumption values; escapism and aesthetics in particular lead to visitor satisfaction (Lee et al., 2017). Attendance at the EPCOT festivals leads to visitors participating in a complex, multi-faceted cycle of consumption.

EPCOT and Its Four Festivals – A Case Study

EPCOT has been a subject of fascination for scholars over the decades. Originally envisioned by Walt Disney to be a city with utopian qualities, it was intended for people to interact in a "beautiful, comfortable, and inspiring public setting" (Gennawey, 2011, p. 366). EPCOT was to take theme parks beyond fantasy leisure spaces to an experiment in urban planning with a resident population and conceptions of the values of communities (Clavé, 2007). This plan was changed when Walt Disney passed away in 1966, and it would instead become a theme park with a future-focused area and a country showcase, in essence a permanent world expo.

Though Disney has developed a couple of other theme park models now, EPCOT was the first new concept for them after the Disneyland/Magic Kingdom paradigm that has now proven so successful around the world. EPCOT would be overtly educational. The clearest explanation of the novel concept was provided by the late Walt Disney Imagineer, Marty Sklar. Sklar (as cited in Beard, 1982) described EPCOT as "a permanent world's fair of imagination, discovery, education, and exploration that combines the Disney entertainment and communications skills with the knowledge and predictions for the future of authorities from industry, the academic world, and the professions" (p. 15). The lofty goal of the park was to educate and inspire through its two sides, Future World and World Showcase. The park is a combination of the "technological future" and the "traditional cultural past" (Telotte, 2008); though this is a seemingly strange combination, it is a long-term staple of the world expo.

In line with the edutainment agenda of Disney, EPCOT had a primary goal of communicating information to the public in an approachable way. It overtly carried a message of science and industry meeting humanity's needs, with a strong message of capitalism through usage of corporate sponsors (Sklar, 2013). Education can be skewed with corporate interests, as Sklar (2013) mentioned happened with the Universe of Energy/Ellen's Energy Adventure attraction, where Disney had final approval over messaging but had competing interests from sponsor Exxon and government energy officials. Even with the interests of corporate sponsors, EPCOT's Future World is a space that communicates science (Martin, 2019). Like in their educational films, the parks have also been sites with "interwoven documentary and narrative elements" (Van Riper, 2011, p. 4), with visitors more likely to want to connect with seemingly dull subjects (e.g. transportation, energy, farming, medicine) if the attraction is fun in addition to didactic.

World Showcase, the impetus for the festival concept, was another nod to the world's fairs, with country pavilions usually created in concert with national governments. Eleven countries made the cut: Mexico, Norway, China, Germany, Italy, United States of America, Japan, Morocco, France, United Kingdom, and Canada. Countries considered but that did not make it include an African nations pavilion, Costa Rica, Denmark, Iran, Israel, Spain, Switzerland, and Venezuela (Sklar, 2013). In this large area of the park, buildings and landscapes signify countries (Bryman, 1995), forming a kind of shorthand to the selected national cultures. The restaurants add to the thematic landscape of the park, emphasising the atmosphere and cuisine of a represented culture (Clavé, 2007). Though EPCOT was originally focused on technological application, the world cultures part of the park has now taken on the "central role" (Thomas, 2011).

EPCOT Festivals

In the 1990s, EPCOT was "struggling" with its image as an adult attraction, finding that the park was the most likely to be left out of a vacation, with repeat visitation lower than other parks (Byrd, 1996). Thus, the park strategically added more attractions for children and thrill seekers while also becoming a "magnet for locals", the park's primary market, through the addition of four festivals (Byrd, 1996). Attendance tended to dwindle in fall, but culinary festivals had worked at several hotels as well as at the Food & Wine Classic in Aspen, Colorado (Mumpower, 2016). Moved from Disney's shopping and dining district to EPCOT, the second of the four festivals, the International Food & Wine Festival, began in 1996 to great popularity. By its third year, it was already the largest festival of its type in the world (O'Brien, 1998). By its tenth year, it was setting revenue and attendance records; it had grown by more than 700% and gained a big following (O'Brien, 2002).

The festivals (see Table 4.1) altered the image of Walt Disney World, going beyond stereotypical amusement park fare and extending into the fine dining and wine connoisseur spaces (Ewing-Mulligan & McCarthy, 2005). In addition

Table 4.1 Details and characteristics of the four EPCOT festivals

	International Festival of the Arts	International Flower & Garden Festival	International Food & Wine Festival	International Festival of the Holidays
Season	Mid-January to Late February	Early March to Early July	Mid-July to mid-November	Thanksgiving Week – New Year's Eve
Began	2017	1994	1996	1996
Signature Elements	Art, music, food, and beverage	Gardens, topiaries, food, and beverage	Food and beverage	Storytelling, food, and beverage
Demonstrations	Artists	Gardeners	Chefs, sommeliers	Storytellers
Concert Series	Disney on Broadway Concert Series	Flower Power Concerts (until 2015); Garden Rocks Concerts Series (2016–now)	Eat to the Beat Concert Series	Candlelight Processional
Scavenger Hunt	Figment's Brush with the Masters	Spike's Pollen-Nation Exploration	Remy's Ratatouille Hide & Squeak	Olaf's Holiday Tradition Expedition
Non-Country Food and Beverage Themes	*Art* ("Artist's Table", "Craftsman's Courtyard", "Deco Delights", "Gourmet Landscapes", "Pastoral Palate")	*Farm fresh* ("Farmer's Feast", "Farmer's Market")*Specific edibles* ("Citrus Blossom", "Honey Bee-stro", "Pineapple Promenade")*Plant-based* ("Trowel & Trellis")	*Specific edibles* ("Cheese Studio", "The Fry Basket", "Great Beers of the World")*Plant-based* ("Earth Eats", "Terra")	*Holiday-based* ("Feast of the Three Kings", "L'Chaim! Holiday Kitchen")*Sweets* ("Chestnuts and Good Cheer", "Holiday Hearth Desserts")

to improving the perception of theme park fare for food lovers (Mintz, 1998), the food started to become an attraction itself and a major draw for visitors. Though planned for repeat visitation by locals, Walt Disney World has domestic visitors as well as millions of foreign tourists, particularly from the largest international markets: Canada, the United Kingdom, Brazil, and Argentina (EDR, 2021).

Research Design and Data Collection

This chapter draws on years of in-person experience and participant observation at EPCOT festivals, with a few visits from 2000 to 2001, sporadic visits from 2006 to 2008, and frequent visits from 2010 to 2022, when the author has had an annual pass, totalling more than 125 total visits. The author was an overt participant with "insider positionality" and would be classified as an "intimately involved insider" (Radel, 2018, p. 134) being a frequent theme park guest, local, annual passholder, themed entertainment researcher, and former Disney theme park employee. While situatedness is useful for site knowledge and understanding of practices, there are clear limitations in that there is bias by frequency and further removed from regular guests. Having full membership in a group while also utilising an objective research presence and theoretical foundation leads to a tricky navigation (Radel, 2018). Nonetheless, the work is couched in the descriptive interpretivist paradigm, wherein individual experience and interpretation are paramount and details, rather than generalisability, are the purpose (Rossman & Rallis, 2016).

In addition to participation, the festival guides and maps for each festival from nearly every year were collected to track the content and evolution of festivals over time. Articles and photographs on blogs were examined for the years prior to the author's experiences. As the focus was solely on the author's personal participating experience rather than interviewing or formally observing park guests, no ethical clearance was needed. This work also attempted to find patterns of activities based on this subjective experience. Festival activities were divided into functions by the use of descriptive coding, and areas where there were simultaneous codes were grouped into their primary function (Saldaña, 2009). Inductive coding derived from the researcher's experience was utilised to generate themes and then analysed with reference to previous literature (Fereday & Muir-Cochran, 2006). Future research could gather additional qualitative data through interviews or operationalise with a quantitative survey to determine if additional themes emerge.

Results and Discussion

The four categories derived from observation related to how the festivals function: as revenue strategy, spaces for play, sites for cultural immersion, and tools for education. Their dispositions in each category are described in Figure 4.1.

Revenue Strategy	Play Experience	Cultural Immersion Site	Educational Tool
Used as a revenue source with guest purchases and exhibitor fees	Utilised to encourage family play	Implemented to provide awareness of global cultures	Employed to instruct guests
Enacted through corporate sponsors, merchandise lines, extra fee experiences	Enacted through interactive quests and scavenger hunts in physical space and other participatory activities	Enacted through food and beverage stands, live performers, exhibits, and designed spaces	Enacted through demonstrations, seminars, exhibits, and signage

Figure 4.1 Four functions of EPCOT festivals and how they are enacted.

Festivals as Play Experiences

While a couple of the core themes of the festivals cater to adults, all the events include activities for children and families. Several have playgrounds and children's activities such as gardening sites or art installations that kids can interact with. During the Food & Wine Festival, a Junior Chef programme that lasted for years allowed children to make cookies and other foods. One of the most prominent activities of each festival is the scavenger hunt (see Figure 4.2), an activity that has become popular with children and adults alike. They all

Figure 4.2 The butterfly garden, a stop on the scavenger hunt Spike's Pollen-Nation Exploration, during EPCOT's International Flower & Garden Festival.
Photograph by the author.

involve following clues or looking around in particular areas to spot a character, for instance, Figment (the park's mascot, from the ride Journey into Imagination with Figment) during the Festival of the Arts, Spike (a character from early Disney animated cartoons) in the Flower & Garden Festival, Remy (from the 2007 animated Pixar film *Ratatouille*) during the Food & Wine Festival, and Olaf (from the 2013 animated Disney film *Frozen*) in the Festival of the Holidays. In each quest, the player finds the item in the quest, learns about a topic (art pieces, flowers, ingredients, etc.), places a sticker on the map, and then gains a small prize for completion. During the festivals, the quests become some of the most physical activities at Disney, as the theme park is large at 300 acres and clues are hidden all over.

These hunts draw on the long-time family activities of the World Showcase passport with stamps and Kidcot Fun Stops, locations in each country that offer stick puppet colouring opportunities with ambassadors from the represented nations. The nature and quality of family play have been altered by EPCOT's space-based experiences that encourage physical activity and problem solving (Eddy et al., 2020; Newman, 2015), and the festivals spawned a new culture of quarterly participation. The collective aspect of so many people participating in these experiences together, in addition to fan groups with loyalty to the festivals and quests, promotes the notion of community. It has been found that some festivals do not have particularly diverse attendees and that bonding with one's own group is a greater value at a festival than bridging with unconnected visitors (Wilks, 2011). Nonetheless, Walt Disney World has a diverse mix of visitors particularly in nationality and domestic regions, though presumably less so in socioeconomic status (EDR, 2021). There also remains a dedicated group of fans that visit the parks and connect both in person and online (Williams, 2020).

While some criticism of World Showcase pavilions focuses on them as static sets instead of living, vibrant places like the actual countries being represented, the quests and scavenger hunts create dynamic scenes. Some clues end up being well hidden, allowing for in-depth discovery and more time in each country not necessarily shopping or eating. Additional time spent in each pavilion during festivals allows guests to be more likely present for a live performance or a closer look at the architectural elements of the space. Aesthetics play a major role in the construction of perceived authenticity at EPCOT (Houston & Meamber, 2011). While the Flower & Garden Festival is especially beautiful with its widespread gardens and topiaries and the Festival of the Arts emphasises colourful arts and crafts, all four festivals allow for exploration and play, which encourage more in-depth exploration of the park's environs. Guest satisfaction in a theme park relies on both educational and aesthetic experiences, but revisit intention is most tied to the concept of escapism (Lee et al., 2020), which these quests are likely to facilitate. The EPCOT festival cycle has permanently impacted the built environment of the theme park (Getz et al., 2018). EPCOT is now a permanent festival space that altered existing structures and added a new layer of structures to hold festival shops and restaurants.

Festivals as Cultural Immersion Sites

Considering the land's existence as a cultural awareness centre, World Show-case is the site of most of the festival activities and there is a lot of cultural content in it. Learning about and immersing in world cultures through their people, customs, and products is a core element of the festival cycle as well as the park itself. Many of the pavilion employees are from the Cultural Repre-sentatives programme, where workers from each nation are recruited to repre-sent their countries at EPCOT. Festivals highlight food, beverage, crafts, music, and other cultural outputs. The Festival of the Holidays showcases cultures most directly, as it includes live storytellers and written signs that discuss the holiday traditions of the World Showcase countries. It also includes the Candlelight Processional, a narrated, musical celebration of the life of Jesus Christ, the most overtly religious offering at a Disney theme park. Though often utilising the concepts of "staged authenticity" and realism rather than objective authenticity and exact portrayal of nations (Milman, 2013), the areas use many techniques to appeal to visitors' sense of cultural discovery.

One impact of the festival cycle is the ability to extend the presentation to cultures outside of the 11 primary countries. The park's Africa area is embar-rassingly represented by a merchandise shop and Coke kiosk. During the festivals, however, Africa has been represented by authentic food from vari-ous regions. Over the years, many countries and regions beyond the core pavilions have been represented in the festivals including Africa, the Alps, Argentina, Australia, Austria, Belgium, Brazil, the Caribbean, Chile, Domin-ican Republic, Greece, India, Ireland, Kenya, Mediterranean, New Zealand, Patagonia, Poland, Peru, Puerto Rico, Scandinavia, Singapore, South Africa, South Korea, Spain, Switzerland, Thailand, Turkey, and US regions. Coun-tries are illustrated as less homogenous with the festivals. For instance, the food in the USA pavilion was burgers and fries and then barbeque. Through-out the years of the festival, though, several cuisines have been showcased, including California, Florida, Hawaii, New England, New Orleans, and Oklahoma. Likewise, other existing pavilions have focused on specific regions within the country (e.g. Yukon in Canada, Shanghai in China, and Hokkaido in Japan).

There has been a sense that guests are either more likely to want to visit the countries represented or less likely (Blyth, 2020). It is "simulated tourism" rather than actual tourism (Mintz, 1998), but several countries can be traversed in one day in an easy-to-accomplish loop. One study (Milman, 2013) explained that guests find the World Showcase pavilions authentic and were more likely to find the content realistic if they had visited the original countries. Another work (Firat & Ulusoy, 2011) found that visitors do not tend to distinguish between fantasy and reality at EPCOT, even those who have visited the original countries. Part of this is due to the environmental design present in theme parks as well as the core of edutainment, where learning is not as in-depth as in some contexts but can stimulate engagement (Kerz, 2016). The park has been criticised for Eurocentrism, Western-centric, or American-focused attractions (Hodge, 2018; Nelson, 1986; Samman, 2010), with only a few outlier pavilions. However, this can be complicated through utilising designers from that culture to foster authenticity (Sheppard, 2016) or through the festivals which have introduced many different cultures to the mix and have helped to combat the notion of its lack of diversity.

Festivals as Educational Tools

EPCOT opened as the most obviously educational Disney park, with only Disney's Animal Kingdom a contender today. The educational orientation and "original vision" of EPCOT to "present educational opportunities for guests" through its pavilions has remained a frequent subject of discussion in fan communities (Williams, 2020, p. 221). The park had an initial purpose to be both entertaining and enriching (Newman, 2015), and that goal continues today, seen perhaps most obviously with the festivals.

The festivals themselves have always had a component of education. Food & Wine has had cooking demonstrations, wine tastings, seminars, signs, booklets, and other educational content from the beginning. A Walt Disney World publicist stated that the wine seminars were popular and a "good way for people to learn about wines" (O'Brien, 1998). An event operations manager mentioned that the festival offerings, with small portions especially, make for a low-risk way to try foods; she continued that it was a "great education in wine and foods" (O'Brien, 2002). Unlike the more exclusive events such as the aforementioned food festival in Aspen, Disney's festivals have a low barrier to entry but are very scalable. A family can just do a scavenger hunt and eat at festival marketplaces, or they could go to seminars and demonstrations or do the more premier experiences.

It is possible that the art festival can create impactful art experiences as well, with its fun ways to engage with creativity from high art to pop art (see Figure 4.3). Considering the subjects of flowers and holidays, the other two festivals are even more accessible because of their less complex themes. Learning about gardening while experiencing different ways to garden might spark behavioural change, while reading about or listening to the holiday practices from around the world can help someone gain appreciation and openness. Even if a powerful transformation does not occur in many of the guests that attend, there is still a compelling layer of meaning in each festival's insistence on educating while entertaining. Over the years, the festivals' continuing focus on education has allowed an orientation towards absorbing information and cultural immersion and the possibility at least of transformation, as learning in festivals can lead to intention to alter behaviour (Ibrahim et al., 2015). There are individuals who find value in learning about places, society and culture, and the arts that take place during the festivals (Getz et al., 2018).

Figure 4.3 An educational sign next to a photo opportunity during EPCOT's International Festival of the Arts.

Photograph by the author.

Festivals as Revenue Strategy

In addition to enhancing visitors' learning and cultural awareness, the four festivals also offer the benefit of revenue generation, which, although not directly related to edutainment, allows the festivals to continue each year and thus enables future edutainment experiences. The four festivals have several core attributes that tie into revenue. They have a vast array of corporate sponsors that create, for example, specialised merchandise lines, extra-fee experiences, and "tapas" style small portions – all to encourage additional edutainment consumption. These festivals are thus sought after by locals because of special events, concert series, prizes for filling in food quests, and desire to taste multiple items, all of which ensure repeat visitation. To this end, for many years Walt Disney World sold an "EPCOT After 4" pass, which specifically existed for locals – from foodies who want to sample different restaurants each week, to the popular phenomenon of "drinking around the world". While food and beverage or merchandise are traditionally seen as ancillary revenue streams, they form a key attraction for visitors during festivals, turning them into a central aspect of the edutainment experience.

Beyond benefit to Disney, the associated organisations and businesses view their festival participation as a chance for increased brand awareness and revenue. For example, when the Food & Wine Festival showcased products and culture from Oklahoma, the state's governor thought that the event offered a chance not only to expose visitors to the beauty of the state (i.e. helping them to learn about Oklahoma and its people) but to promote tourism and stimulate economic development opportunities (Oklahoma Gazette, 2007). Businesses thus primarily attend food festivals for economic and marketing reasons. However, they can be motivated by personal interests and passion or the desire to showcase their traditions and culture (Janiszewska & Ossowska, 2021). In this case, the businesses chose to engage with the festivals for the edutainment benefits of the exhibit – to engage beyond their financial motives.

Conclusions

Edutainment, a concept defined by theme park innovator Walt Disney, is showcased in rides and shows throughout the theme park industry. It has now extended to even temporary festivals housed within these permanent spaces. Festivals with an edutainment orientation have proven to be increasingly popular entries into the theme park operational milieu. Several other theme park festivals have been inspired by the success of EPCOT, including the Busch Gardens Food & Wine Festival, Knott's Boysenberry Festival at Knott's Berry Farm, the SeaWorld Seven Seas Food Festival, Universal's Mardi Gras: International Flavors of Carnaval, and the Tastes of Grand Carnivale at Cedar Fair parks. While these events have brought additional attendance and revenue, none compare with the scope or level of depth in the EPCOT festivals.

The EPCOT cycle illustrates that festivals can operate on multiple levels at once, generating profit and increasing attendance while simultaneously offering family play, cultural immersion, and opportunities for learning. Though the annual festival cycle began as a tool to generate revenue and encourage consumption, it has become a source for edutainment and community, recalling the initial lofty visions for the park. With the connections between festivals, learning, and behavioural change, there are wide-reaching impacts. Whether a large-scale annual festival at a theme park with revenue and attendance goals or a small-scale community-building venture without profit motives, festivals that employ techniques of edutainment become more engaging and conducive to learning and meaning. The visit to the festival can then bring change, becoming potentially transformative for individuals and communities.

References

Amorin, D., Jiménez-Cabellero, J. L., & Almeida, P. (2019). Motivation and tourists' loyalty in performing arts festivals: The mediator role of quality and satisfaction. *Enlightening Tourism*, *9*(2), 100–136. https://doi.org/10.33776/et.v9i2.3626

Ascione, E., & Fink, C. (2021). Italian sagre: Preserving and re-inventing cultural heritage and community through food festivals in Umbria, Italy. *Food, Culture & Society*, *24*(2), 291–308. https://doi.org/10.1080/15528014.2021.1873037

Beard, R. R. (1982). *Walt Disney's EPCOT center: Creating the new world of tomorrow*. Harry N. Abrams, Inc.

Blyth, J. (2020). *Polishing the dragons: Making EPCOT's wonders of China*. Bamboo Forest Publishing.

Bryman, A. (1995). *Disney and his worlds*. Routledge.

Byrd, A. (1996, October 11). Epcot entertains change: Disney World gives 'struggling' park new image. *Orlando Business Journal*. https://link.gale.com/apps/doc/A18964501/GBIB?u=orla57816&sid=bookmark-GBIB&xid=7b17975c

Chang, Y.-C. (2020). Creating value through the performing arts: The multi-stakeholder approach. *Journal of Macromarketing, 40*(2), 185–200. https://doi.org/10.1177/0276146719894627

Choo, H., Park, D.-B., & Petrick, J. F. (2022). Festival tourists' loyalty: The role of involvement in local food festivals. *Journal of Hospitality and Tourism Management, 50*, 57–66. https://doi.org/10.1016/j.jhtm.2021.12.002

Clavé, S. A. (2007). *The global theme park industry*. CABI.

Eddy, R., Baker, C., Macy, R., Murray, J. T., & Salter, A. (2020). Hacking droids and casting spells: Locative augmented reality games and the reimagining of the theme park. *Proceedings 31st ACM conference on hypertext and social media*, 37–46, Virtual Event, USA.

EDR (Office of Economic & Demographic Research). (2021). *Return on investment for Visit Florida*. The Florida Legislature.

Ewing-Mulligan, M., & McCarthy, E. (2005). Walt Disney World shows its grown-up side with vast, varied wining and dining. *Nation's Restaurant News. 32*.

Fereday, F., & Muir-Cochrane, E. (2006). Demonstrating rigor using thematic analysis: A hybrid approach of inductive and deductive coding and theme development. *International Journal of Qualitative Methods, 5*(1), 80–92. https://doi.org/10.1177/160940690600500107

Fırat, A. F. & Ulusoy, E. (2011). Living a theme. *Consumption Markets & Culture, 14*(2), 193–202. https://doi.org/10.1080/10253866.2011.562020

Gennawey, S. (2011). *Walt and the promise of progress city*. Ayefour Publishing.

Getz, D., Andersson, T. D., Armbrecht, J., & Lundberg, E. (2018). The value of festivals. In J. Mair (Ed.), *The Routledge handbook of festivals* (pp. 22–30). Routledge.

Hodge, M. (2018). Disney "World": The Westernization of world music in EPCOT's "Illuminations: Reflections of Earth". *Social Sciences, 7*. https://doi.org/10.3390/socsci7080136

Houston, H. R., & Meamber, L. A. (2011). Consuming the "world": Reflexivity, aesthetics, and authenticity at Disney World's EPCOT Center. *Consumption Markets & Culture, 14*(2), 177–191. https://doi.org/10.1080/10253866.2011.562019

Ibrahim, N., Ismail, N. A., Yunos, M. Y. M., Utaberta, N., Ariffin, N. F. M., & Ismail, S. (2015). Evaluating the effectiveness of garden festival in educating environmental design behaviour through Royal Fluria Putrayaja Flower and Garden Festival. *Advances in Environmental Biology, 9*(24), 121–126.

Janiszewska, D., & Ossowska, L. (2021). Food festival exhibitors' business motivation. *Sustainability, 13*. https://doi.org/10.3390/su13094920

Kerz, C. (2016). Atmosphere, immersion, and authenticity in Colonial Williamsburg. In S. Lukas (Ed.), *A reader in themed and immersive spaces* (pp. 195–205). ETC Press.

Kirshenblatt-Gimblett, B. (1998). *Destination culture: Tourism, museums, and heritage*. University of California Press.

Knight, C. K. (2014). *Power and paradise in Walt Disney's World*. University Press of Florida.

Kratz, C. A., & Karp, I. (1993). Wonder and worth: Disney museums in World Show-case. *Museum Anthropology*, *17*(3), 32–42. https://doi.org/10.1525/MUA.1993.17.3.32

Lee, I., & Arcodia, C. (2011). The role of regional food festivals for destination brand-ing. *International Journal of Tourism Research*, *13*, 355–367. https://doi.org/10.1002/jtr.852

Lee, S., Jeong, E., & Qu, K. (2020). Exploring theme park visitors' experience on satis-faction and revisit intention: A utilization of experience economy model. *Journal of Quality Assurance in Hospitality & Tourism*, *21*(4), 474–497. https://doi.org/10.1080/1528008X.2019.1691702

Lee, W., Sung, H., Suh, E., & Zhao, J. (2017). The effects of festival attendees' experien-tial values and satisfaction on re-visit intention to the destination. *International Jour-nal of Contemporary Hospitality Management*, *29*(3), 1005–1027. https://doi.org/10.1108/IJCHM-10-2015-0559

Leer, J., & Juel-Jacobsen, L. G. (2022). Food festival experiences from visitors' perspec-tives: Intellectual, sensory, and social dimensions. *Food and Foodways*, *30*(4), 287–309. https://doi.org/10.1080/07409710.2022.2124729

Liang, Z., & Li, X. (2021). What is a theme park? A synthesis and research framework. *Journal of Hospitality & Tourism Research*. https://doi.org/10.1177/10963480211069173

Mair, J., & Duffy, M. (2018). The role of festivals in strengthening social capital in rural communities. *Event Management*, *22*, 875–889. https://doi.org/10.3727/152599518X15346132863229

Martin, D. (2019). EPCOT theme park as science communication space: The Test Track case. *Journal of Science Communication*, *18*(4), 1–12. https://doi.org/10.22323/2.18040209

Martin, G. (2022). Here's why Disney fans lined up for hours to buy a popcorn bucket. *Paste Magazine*.

Milman, A. (2013). Guests' perceptions of staged authenticity in a theme park: An example from Disney's Epcot's World Showcase. *Tourism Review*, *68*(4), 71–88.

Mintz, L. (1998). Simulated tourism at Busch Gardens: The Old Country and Disney's World Showcase, Epcot Center. *Journal of Popular Culture*, *32*(3), 47–58. https://doi.org/10.1111/j.0022-3840.1998.3203_47.x

Misener, L. (2013). Events and social capital. In R. Finkel, D. McGillivray, D. McPherson, & P. Robinson (Eds.), *Research themes for events* (pp. 18–30). CABI.

Mumpower, D. (2016). History of Epcot international food and wine festival at Walt Disney World. *DVC Resale Market Blog*. DVC Resale Market.

Naccarato, P., & LeBesco, K. (2012). *Culinary capital*. Berg Publishers.

Nelson, S. (1986). Walt Disney's EPCOT and the world's fair performance tradition. *The Drama Review*, *30*(4), 106–146. https://doi.org/10.2307/1145786

Newman, J. (2015). The future of family play at Epcot. In M. Omasta & D. Chappell (Eds.), *Play, performance, and identity: How institutions structure ludic spaces* (pp. 55–66). Routledge.

O'Brien, T. (1998). Wine & food tasted at Epcot festival. *Amusement Business*, *110*(30), 56.

O'Brien, T. (2002). Epcot's food and wine festival has grown by 700% since its 1996 start. *Amusement Business*, *114*(46), 14.

Oklahoma Gazette. (2007). *Gov. Henry, Disney officials announce Epcot festival spot-lighting Oklahoma*. https://www.okgazette.com/oklahoma/henry-disney-officials-make-centennial-epcot-announcement/Content?oid=2958156

Organ, K., Koenig-Lewis, N., Palmer, A., & Probert, J. (2015). Festivals as agents for behaviour change: A study of food festival engagement and subsequent food choices. *Tourism Management*, *48*, 84–99. https://doi.org/10.1016/j.tourman.2014.10.021

Putnam, R. D. (1995). Bowling alone: America's declining social capital. *Journal of Democracy*, *6*(1), 65–78.

Radel, K. (2018). Participant observation in cross-cultural tourism research. In W. Hillman & K. Radel (Eds.), *Qualitative methods in tourism research: Theory and practice* (pp. 129–156). Channel View Publications.

Ritzer, G. (2010). *Enchanting a disenchanted world: Continuity and change in the cathedrals of consumption*. Pine Forge Press.

Rossman, G. B., & Rallis, S. F. (2016). *An introduction to qualitative research: Learning in the field*. SAGE.

Rubin, J. (2020). *Global attractions attendance report 2019*. TEA/AECOM.

Rubin, J. (2022). *Global attractions attendance report 2021*. TEA/AECOM.

Saldaña, J. (2009). *The coding manual for qualitative researchers*. SAGE.

Samman, K. (2010). The temporal template of tourism: A comparative analysis of Epcot Center (Orlando) and Wadi Rum (Jordan). *Journal of Tourism and Cultural Change*, *8*(4), 305–315.

Sheppard, R. (2016). Mexico goes to Disney World: Recognizing and representing Mexico at EPCOT Center's Mexico pavilion. *Latin American Research Review*, *51*(3), 64–84.

Sklar, M. (2013). *Dream it! Do it!: My half-century creating Disney's Magic Kingdoms*. Disney Editions.

Telotte, J. P. (2008). *The mouse machine: Disney and technology*. University of Illinois Press.

Thomas, G. (2011). Walt Disney's EPCOT: The city of tomorrow that might have been. *The Objective Standard*, *6*(1), 103.

Tzaczyski, A., Knox, K., & Rundle-Thiele, S. (2022). A small-scale festival as a catalyst for individual and community change. *Event Management*, *26*, 1833–1848. https://doi.org/10.3727/152599522x16419948391249

Van Riper, A. B. (2011). Introduction. In A. B. Van Riper (Ed.), *Learning from Mickey, Donald and Walt: Essays on Disney's edutainment films* (pp. 1–14). McFarland.

Wei, W. (2018). Understanding values of souvenir purchase in the contemporary Chinese culture: A case of Shanghai Disney. *Journal of Destination Marketing & Management*, *10*, 36–48.

Wilks, L. (2011). Bridging and bonding: Social capital at music festivals. *Journal of Policy Research in Tourism, Leisure & Events*, *3*(3), 281–297.

Williams, R. (2020). *Theme park fandom: Spatial transmedia, materiality and participatory cultures*. Amsterdam University Press.

Wright, A. (2006). *The Imagineering field guide to Epcot*. Disney Editions.

Wyatt, B., Leask, A., & Barron, P. (2021). Designing dark tourism experiences: An exploration of edutainment interpretation at lighter dark visitor attractions. *Journal of Heritage Tourism*, *16*(4), 433–449. https://doi.org/10.1080/1743873X.2020.1858087

5   SouthWestFest

Edutainment through the Lens of a Cultural Community Festival, and Its Delivery Partners

Joanna Goodey

Introduction

SouthWestFest is an award-winning cultural community festival that takes place each year in the city of Westminster, in inner London. The festival works with partner organisations to deliver events that offer positive social outcomes to residents and organisations. A particular feature of the festival is that it includes opportunities for learning; these are delivered as part of its overall entertainment offering.

This chapter will investigate how educational activities delivered by the festival, in conjunction with its partners, may be understood as forms of edutainment. Edutainment can be defined as the enhancement of learning when facilitated through education delivered in conjunction with entertainment. This is then done with the intention to attract and excite the learner by making the learning experience more enjoyable (Aksakal, 2015). To explore this further, the chapter presents a case study of festival activities which have been produced in a collaboration between SouthWestFest and three of its partner organisations: Westminster Abbey, City Lions, and Pimlico Toy Library. By reflecting on this work, a range of practitioner perspectives are presented. In particular, planning and design choices are investigated in order to understand how the combination of education and entertainment can lead to learning outcomes being delivered and experienced through and within the environment of a community festival.

Edutainment and the 'festivalised' servicescape

An early example of edutainment was Walt Disney's successful Factual Nature Pictures series, which, from 1954 onwards, were widely used as scholarly aids in schools in the United States (Disney, 1954). The use of edutainment as a tool for learning both inside and outside of formal educational settings grows year on year (Jarvin, 2015). It continues to be associated with film, but now also encompasses other realms of media entertainment, such as TV, music, computer software and games. More recently, festivals and events have also provided sites of entertainment and learning, allowing opportunities for both planned and organic forms of edutainment to take place (d'Hauteserre, 2011).

DOI: 10.4324/9781003305415-6

Recognition of edutainment's facility for enhancing the educational, experiential, and positive social outcomes of events has led to edutainment being researched alongside other event impacts and outcomes. Science and literature festivals offer clear potential for linking entertainment and learning (Jensen & Buckley, 2012; Rossetti & Quinn, 2019), whilst the role and outcomes of edutainment for other types of festivals such as food festivals (Carvalho et al., 2021), arts festivals (Viljoen et al., 2018), film festivals (Herrschner, 2015), and music festivals (Sisson & Alcorn, 2022) have also been investigated. Although some scholars have begun to note edutainment as a potentially important outcome of local cultural events (Kim, 2017), scant research has been undertaken to date regarding forms and outcomes of edutainment that are directly related to community festivals and to the importance of both festival's setting.

The popularisation of festivals across all sectors and artistic genres is reflective of the general rise in consumer demand, particularly amongst the young, for ephemeral experiences rather than for products, a trend that has given rise to the 'experience economy'. Gilmore and Pine (1999) suggested that services that embody an overlap of the four realms of experience design – entertainment, education, aesthetics, and escapism – provide opportunities for both interactive and passive forms of participation. Intentional curation of the festival experience can also be useful for the marketing of event experiences, with the notion that event programmes which become 'festivalised' offer a more exciting incentive for audiences (Négrier, 2015).

Festivals and events can be considered as a service environment or servicescape (Bitner, 2000), deliberately designed to deliver tangible effects on consumer behaviour, favourable to the festival organiser's desired outcomes. Moreover, numerous studies (Bruwer, 2012; Lee et al., 2008; Yoon et al., 2010) now draw on the concept of the festivalscape, identifying both the tangible and intangible cues specific to a festival environment which lead to the intentional shaping of the festival consumer experience to evoke a positive emotional response and enhance overall visitor satisfaction. Although these studies generally refer to the curation of the festivalscape as being helpful to the perceived service quality and economic value of the festival, Chou et al. (2018) and Chen et al. (2019), amongst others, have argued that the cues which shape the consumer's experience of the festival environment can also support wider positive outcomes. Indeed, community festivals have long been heralded for the wide range of intangible and tangible benefits they offer to their host communities and stakeholders, including showcasing talent, promoting culture and heritage, providing informal learning experiences (Paris, 2002), developing both social and cultural capital for attendees, in addition to developing social sustainability through improving wellbeing (Brownett & Evans, 2019; Stevenson, 2020), as well as supporting a wider sense of belonging, community cohesion, and place-making (Andrews & Leopold, 2013).

Community festivals have also been identified as sites conducive to forms of public pedagogy, wherein autonomous community-led learning takes place through participation in a festival setting in a public space (Li et al., 2017).

The forms of learning that take place within these festival sites might be considered 'non-formal learning', which Rogers (2014, p. 12) defines as learning that takes place through a planned educational activity, outside of the formal educational setting, or alternatively, 'informal learning', defined by McGivney (2002, p. 1) in which people accumulate understanding, skills, and knowledge from their everyday life. Learning which takes place at festivals and events can also be considered experiential learning, where learning is undertaken through embodied experience and strengthened through memory making (Kolb, 2015). As such, it is interesting to further consider how community festivals and their partners approach the design and curation of the festival experience and entertainment to support or enhance non-formal or informal learning opportunities, as forms of edutainment.

Methods

This chapter sets out to consider how the specific cues, environment, and setting of the SouthWestFest festivalscape are utilised by the festival and its partners concerning the planning and design of learning and engagement activities which can take place in conjunction with entertainment. It does this by offering an interpretive study of practitioner perspectives on approaches to learning design and production of educational activities delivered for and in collaboration with SouthWestFest festival, identifying outcomes of learning as edutainment.

The chapter uses a qualitative research methodology to support the investigation of the choices, attitudes, and behaviours of those working within leisure settings such as community festivals (Veal, 2017). The method of a case study was chosen to offer a holistic investigation in order to consider the complexities of the relationships between the festival and its partners, within the wider context of real-life events in the community festival setting (Leavy, 2014). Non-probability sampling methods were used in support of the small exploratory research for this chapter. Interviews were conducted with SouthWestFest alongside three of the festivals partner organisations, which were selected as a purposive sample to provide a range of experience (in years) of working with the festival, as well as offering a variety of services and activities designed for different subsets of audiences (family, youth, children) and learning outcomes. This method was chosen to support the investigation of a range of experiences and offer specific insights from different practitioners' expertise.

Research for the case study presented includes primary data from four interviews which were held with a member of staff from the festival itself and a member of staff from each of the partner organisations; this was supported by secondary research from websites, organisational reports, and other online sources. Ethics approval for all data collected was sought and approved. Semi-structured, open-ended interview questions were chosen to let the interviewee lead the discussion and reduce the opportunity for bias or coercion (Grindsted, 2005). Interviews were then coded and thematically analysed

(Braun & Clarke, 2022), to form the basis of the case study and key emerging themes highlighted in the discussion presented. It is also important to note the author's close position concerning the research and discussion within the chapter. The author of this chapter was previously the Festival Director of SouthWestFest from 2015 to 2021 and therefore draws on their prior knowledge of festival activities and partnerships.

Learning through entertainment at SouthWestFest – A case study

SouthWestFest is a charity-led cultural community festival which has been held annually for the last 19 years. It takes place each July delivering around 100 free or low-cost events and activities; in addition, in recent years, it has also delivered a series of online events. The festival is supported by more than 70 partner organisations and groups who contribute events, venues, resources, staff, and volunteers to aid the festival programme. Each year the festival is launched with the outdoor Festival Day, which attracts an audience of thousands and has become a hallmark event for the community. Following this, a three-week themed festival programme is delivered across south Westminster. The SouthWestFest festival programme is made up of a curated programme produced by the festival which runs alongside an open access programme which promotes events that are independently produced by local organisations and groups. The festival team also picked a theme, to tie together both sides of the programme (SouthWestFest, 2022).

For the festival team, learning is an integral part of the festival experience, as they explain,

> all our festival activities include learning as a natural part of the experience which can be derived from that event. For example, audiences attending Festival Day might learn about a new style of music or cultural dance. Alternatively, the performers participating will often be part of a performing school, so learning takes place as part of their own self-development.

However, the team also note that it's not just about what's on the stage, as they deliver a range of experiences through both indoor and outdoor events, which are not just entertaining, or cultural experiences, but also embedded into other parts of people's lives such as their wellbeing, education, and careers. They go on to note, "the festival only happens once a year, so can often provide opportunities for unique educational moments, such as learning how to rock climb at one of our outdoor events or writing poetry as part of a heritage trail".

Given that many of SouthWestFest's partner organisations serve cultural, arts, heritage, health, and education sectors, a considerable part of the activities that take place within the festival setting are planned and led by learning professionals; the aim is that the learning outcomes derived from taking part in the festival link back to the educational output of the partner organisation. This aim, in turn, connects to one of the festival's wider charitable objectives: namely, to improve the lives and wellbeing of South Westminster residents by boosting engagement and longer-term interaction with local services (South WestFest, 2020).

A good example of one of SouthWestFest's longest-running partnerships is their work with Pimlico Toy Library, a charity that supports families by offering play sessions, which include toys, play, art and crafts, plus special activities such as day trips, and school holidays, as well as toy loans from a library of 1,500 multicultural and age-appropriate toys (Pimlico Toy Library, 2014). Pimlico Toy Library has taken part in SouthWestFest since its first edition in 2004, with activities including festival parade floats, costume-making workshops, giant games, safe play areas, and arts and crafts. The festival team notes, "Working with Pimlico Toy Library, enables us to design and deliver relevant educational activities for babies, toddlers, young children and their families which are age-appropriate, and promote stimulating and engaging interactions within the festival setting".

To support this, Pimlico Toy Libraries staff draw on their extensive knowledge of early years learning and development theory and best practice. As the Librarians explain, "We look to develop fine and gross motor skills and mental and physical skills. We try to take a creative, innovative and holistic approach". The Librarians also highlighted the importance of factoring in all sorts of opportunities for entertainment into their activities design,

> The barriers to learning are multifold. You've to inspire, motivate and encourage learning. It has to be fun, and entertainment in the widest sense has a huge part to play. So, we're always looking for new experiences to make that point and taking part in SouthWestFest has factored into that.

When considering their interaction with the festival environment, and how it might change or shape learning and engagement activities, they responded,

> The whole atmosphere and environment of the festival is encouraging and uplifting. All the motivation of an upbeat environment enables us to show the children and families we work with that learning is fun, and how important learning is to change things. To prove that children and adults can learn if the environment is right.

The festival themes also help to shape the activities; for example, children and parents will learn about the environment, because they have taken part in a recycled materials costume-making workshop for a sustainability-themed parade. This leads to new and different kinds of learning taking place, as the Librarians explain, "The festival has enabled us to do something a little bit different. It shows parents what their children are capable of, and they are amazed at what they as a family have achieved". The librarians also noted that through responding to the festival theme, the activities take on an element of the festival's entertainment offer, and learning becomes more experiential and relevant, going beyond the regular early years learning.

Each of the festival's partners supports the festival in the creation of unique entertainment experiences, which attract different demographics of audiences. In turn, the partners often want to get involved in the festival, to expand their audiences living in the south Westminster locality. It was through looking to conduct community outreach that the family learning team from Westminster Abbey first connected with SouthWestFest in 2017.

Westminster Abbey is one of the Church of England's most iconic structures and is the location for numerous high-profile national services; in addition, it offers regular daily services and is an extremely important historic, cultural heritage site and popular visitor attraction. An important part of its work is to provide opportunities for learning and engagement in the areas of history, religious education, and culture (Westminster Abbey, n.d.).

One of the most successful collaborations has been Westminster Abbey's contribution to outdoor workshops delivered at the SouthWestFest Festival Day, as the festival team note,

> We need a wide range of content which is appealing to our audiences. Our partners help us with this by leading on activities designed for the workshop zone. Westminster Abbey's team always design engaging and intergenerational activities, in which learning takes place organically. It's interesting to see what will draw people in. Some will stop by for a few moments, and others will stay for over an hour trying out what is on offer.

The Festival Day event offers a uniquely inviting atmosphere for participant engagement, as highlighted by Westminster Abbey's family learning team

> It is a marketplace of opportunities, and people want to try a little bit of everything. You get a lot of tweens to teens who are exploring by themselves because they're in a safe space, on their territory, so they feel like they have complete ownership of their own experience. It's joyful when you have music and an entertaining atmosphere. People are learning, but don't think they are learning because what they are having is a great time, and they are having a great time because learning is fun.

This atmosphere is utilised by the Westminster Abbey team when planning educational activities to deliver with the festival. Mirroring in-house activities, they design activities that are informed by fun, playfulness, song, storytelling, and entertainment, which encourage participants to learn together. However, they adapt the design of activities to ensure they can happen in a shorter time frame to reflect the transient nature of the festival. Activities delivered for the outdoor Festival Day workshops have ranged from Arch building competitions, mirroring the architecture of the Abbey, to dressing up as memorial statues and making scenes for group and family photography portraits,

> these activities create a fun togetherness, help people make happy memories together and at the same time we can have a conversation that leads to learning something about Westminster Abbey, which potentially encourages them to visit us or they might take away leaflets to think or further reflect on.

The Westminster Abbey family learning team went on to reflect upon how taking part in the SouthWestFest Festival Day had informed their in-house approach to learning and events design, commenting,

> We have learnt a lot from the festival on how to welcome people from the area, subconsciously it's informed our family days, which like the festival day, have a marketplace of activity, combined with structured moments of entertainment. We try to recreate the atmosphere that we would have at the festival. It's about creating that experience and memories.

The team also noted that collaborating with and taking part in SouthWestFest supported them in their aim to welcome more local people through the doors of Westminster Abbey, as well as furthering people's learning and understanding of the Abbey's history and existence. Whereas for the festival team, bringing on board prestigious partners such as Westminster Abbey benefits the festival in numerous ways, as they explain "The partnership has helped us develop our cultural and heritage offer, and the rich history of Westminster Abbey inspires people's imaginations when taking part in our festival events".

Like many festivals in 2020, SouthWestFest was forced to flip its delivery to an online festival with major changes in planned provision, as the COVID-19 global pandemic caused local and national lockdown measures across the UK. Previous to these changes, SouthWestFest had been developing a new partnership with City Lions, a youth engagement programme for 13- to 16-year-olds organised by Westminster City Council.

City Lions were already working in Westminster delivering a range of programmes that support young people in their personal development by breaking down the barriers to accessing cultural, and creative sectors, and by acting as a bridge, to encourage young people to explore new interests and opportunities (City Lions, 2022).

Pivoting to online delivery, SouthWestFest delivered its first digital festival and, at the same time, City Lions turned its Creative Industries Week into a youth-based digital festival. Both events ran across the same timeframe, and the organisations came together to produce collaborative events, jointly programmed and promoted across both festivals. City Lions team reflected on this time stating,

> Festivalising the online content made it more attractive, fun and entertaining, working as a bit of a marketing hook. We also had to change our usual programme to include things that were more performative and entertaining, which we found was needed as people adjusted to activities online.

In addition, the festival team highlighted how the collaboration enabled them to develop a better offer for youth-orientated events,

> When lockdown took place, we had to think about how we could reinvent the festival's offer to young people in the area. We hadn't produced online events for young people before and working with City Lions helped us to find new ways to interact with that audience.

As with the other festival partners, the City Lions team recognise that entertainment is a motivational factor in getting young people involved in the programmes they provide. As they pointed out, "Engaging with culture is important because it offers development and learning by stealth". They went on to explain "The young people don't even realise they are learning in the process of being together and having fun". Working together on shared festival events enabled both organisations to mix professional learning development with entertainment, culture, and performance. For example, the two organisations co-produced an online stand-up comedy workshop including a stand-up comedy show. The City Lions team reflected,

> It helps us engage those that are hardest to reach. Attending the performances, the young people might see something of themselves in the people who are performing, and then by taking part realise that actually, this might be for them too.

Further to the online comedy workshop, both organisations collaborated on a creative industries event (Figure 5.1). The festival team added,

> It was a great opportunity to work with and educate younger teens on new potential career opportunities. They may have heard of the festival but not ever considered how it was run. Taking part in the workshop inspired them to consider running their own events within the community.

Finally, both the festival team and City Lions team also recognised other benefits of the collaboration, as shared budgets enabled both organisations to offer more to their respective audiences, and cross-promotion of joint events created opportunities to widen audiences.

Figure 5.1 Festival workshop activities.

Photographer credit: Fabrice Bourgelle, taken for SouthWestFest.

Discussion

Across the case study, similar themes arose within the festival and its partners' responses to planning and design for learning when it takes place outside of their normal spaces and is brought into the festival environment. The majority of learning was shaped towards informal learning outcomes (McGivney, 2002), as the aim for learning within these situations was broad. When thinking about how entertainment factored into learning design, what often came into focus was the terms "fun" and "play", emphasising the emotional and physical experiential reactions to the entertainment, rather than the form of entertainment itself being the motivator for the learning experience. Memory making has long been understood to be an important factor in the success of experiential learning methods (Kolb, 2015), and the partners highlighted this, as they identified the need to create new experiences which supported memory making and which encouraged interactions between individuals learning together.

Like many festivals, SouthWestFest offers a unique one-off experience, which takes place outside of everyday activities, conducive to public forms of pedagogy (Li et al., 2017). SouthWestFest and its partner organisations' design of learning activities were often reshaped, to better fit the 'one-off' transient nature of the festival space and short-term interactions with festival participants. SouthWestFest's outdoor Festival Day was referred to as a marketplace of opportunities, and, consequently, activities needed to be more impactful, to catch people's attention and draw participation. The approaches to learning activities designed for the festival all included elements which echoed Gilmore and Pine's (1999) idea of the 'sweet spot' of experience design. Although this concept was not directly referred to, it is interesting to note the partners' inherent understanding of how expectations of festival participants are shaped within the wider experience economy.

The provision of entertainment alongside learning of activities was beneficial for both SouthWestFest and its partners in widening engagement. The special 'one-off' nature of unexpected or unusual entertainment and festival events acted as a 'hook' to draw in new audiences as participants in informal learning activities. Festivalisation of cultural programmes and activities has long been used as a marketing strategy for audience development (Négrier, 2015). Although not a new idea, the case study still highlighted the added value that the festivalisation of community events and activities can offer, supporting organisations in ensuring sustainable growth in access to their services by residents.

Edutainment at SouthWestFest is delivered and promoted through the festival's collaboration with its partners and can be seen to be taking place through a multitude of direct entertainment formats both predesigned and organic (d'Hauteserre, 2011). These might include on the stage and through workshops, play, arts, and crafts, as well as through the festival participants' immersion within the special celebratory environment that a community festival provides. In the case of the festival and its partners, this offers them the freedom to work differently. The combination of the direct entertainment forms, the entertaining atmosphere, and the experience of both emotional (fun) and physical (play) offer 'stand out' opportunities for the festival and its partners to engage and increase learning. As highlighted in the case study, this then produces learning which happens by stealth, as people do not realise they are learning, because they are less aware of the learning whilst having fun. These experiences become a useful tool to draw in those who might shy away from other educational experiences, or think that learning is not for them, benefiting the festival participants, as well as the festival and its partners.

Conclusion

In concept, edutainment is taking place at SouthWestFest, as entertainment and the environment are utilised in the design and delivery of learning activities. However, interestingly, both the festival organisers and partner organisations interviewed said they had never previously used the word 'edutainment' within their conveying of learning design and strategy to stakeholders or in outward-facing communications to public audiences. This then begs the question, how can further understanding these processes as edutainment be valuable to the festival, its partners, and stakeholders in the future?

There is value in the ability to easily replicate the success factors that derive from edutainment delivered in a community festival atmosphere, into other public spaces and activities as positive learning experiences. In addition, many community festivals and their partners are charity based and rely on external public funding, which often requires stringent measuring of opportunities, outputs, and outcomes. Here, further exploration of the use of the concept of edutainment to support clearer communication of the importance of entertainment as part of learning processes could support festivals and their partners in better evaluation of their objectives and justification of practices and outcomes.

This chapter provides just a small sample of perspectives on how entertainment and education can combine as edutainment to support positive outcomes in a community festival context. Further research into edutainment design in connection to learning approaches and frameworks to support the analysis of its outcomes could be very useful to support the ongoing sustainability of community festivals and their collaborating sectors. In addition, through the ongoing interaction between researchers, festivals, their partners, and practitioners, the concept and terminology of edutainment may be better understood and used for the benefit of all involved.

References

Aksakal, N. (2015). Theoretical view to the approach of the edutainment. *Procedia – Social and Behavioral Sciences*, *186*, 1232–1239. https://doi.org/10.1016/j.sbspro.2015.04.081

Andrews, H., & Leopold, T. (2013). *Events and the social sciences*. Routledge.

Bitner, M. J. (2000). The servicescape. In T. A. Swartz & D. Iacobucci (Eds.), *Handbook of services marketing & management*. Sage Publications.

Braun, V., & Clarke, V. (2022). *Thematic analysis: A practical guide to understanding and doing*. Sage Publications.

Brownett, T., & Evans, O. (2019). Finding common ground: The conception of community arts festivals as spaces for placemaking. *Health & Place*, 102254. https://doi.org/10.1016/j.healthplace.2019.102254

Bruwer, J. (2012). Service quality perception and satisfaction: Buying behaviour prediction in an Australian Festivalscape. *International Journal of Tourism Research*, *16*(1), 76–86. https://doi.org/10.1002/jtr.1901

Carvalho, M., Kastenholz, E., & Carneiro, M. J. (2021). Pairing co-creation with food and wine experiences—A holistic perspective of tourist experiences in Dão, a Portuguese wine region. *Sustainability*, *23*, 13416. https://doi.org/10.3390/su132313416

Chen, Z., King, B., & Suntikul, W. (2019). Festivalscapes and the visitor experience: An application of the stimulus organism response approach. *International Journal of Tourism Research*. https://doi.org/10.1002/jtr.2302

Chou, C. Y., Huang, S.-C., & Mair, J. (2018). A transformative service view on the effects of festivalscapes on local residents' subjective well-being. *Event Management*, *22*(3), 405–422. https://doi.org/10.3727/152599518x15258072560248

City Lions. (2022). *Become a Lion – City Lions*. Retrieved from https://citylions.org/students/

d'Hauteserre, A.-M. (2011). Festival of Pacific Arts: Education in multi-cultural encounters. *International Research in Geographical and Environmental Education*, *4*, 273–280. https://doi.org/10.1080/10382046.2011.619804

Disney, W. (1954). Educational values in factual nature pictures. *Educational Horizons, 33*(2), 82–84. http://www.jstor.org/stable/42922993

Gilmore, J. H., & Pine, J. B. (1999). *The experience economy: Work is theatre & every business a stage.* Harvard Business School Press.

Grindsted, A. (2005). Interactive resources used in semi-structured research interviewing. *Journal of Pragmatics, 7*, 1015–1035. https://doi.org/10.1016/j.pragma.2005.02.011

Herrschner, I. (2015). The role of art in German cultural diplomacy: An analysis of the festival of German films in Melbourne, Australia. *Media Transformations*, 124–141. https://doi.org/10.7220/2029-8668.11.07

Jarvin, L. (2015). Edutainment, games, and the future of education in a digital world. *New Directions for Child and Adolescent Development, 147*, 33–40. https://doi.org/10.1002/cad.20082

Jensen, E., & Buckley, N. (2012). Why people attend science festivals: Interests, motivations and self-reported benefits of public engagement with research. *Public Understanding of Science, 5*, 557–573. https://doi.org/10.1177/0963662512458624

Kim, K. S. (2017). Local festival and culture contents. *Journal of the Korea Convergence Society, 7*, 183–189. https://doi.org/10.15207/JKCS.2017.8.7.183

Kolb, D. A. (2015). *Experiential learning.* Pearson Education.

Leavy, P. (2014). *The Oxford handbook of qualitative research.* Oxford University Press.

Lee, Y.-K., Lee, C.-K., Lee, S.-K., & Babin, B. J. (2008). Festivalscapes and patrons' emotions, satisfaction, and loyalty. *Journal of Business Research, 61*(1), 56–64. https://doi.org/10.1016/j.jbusres.2006.05.009

Li, J., Moore, D., & Smythe, S. (2017). Voices from the "Heart": Understanding a community-engaged festival in Vancouver's Downtown Eastside. *Journal of Contemporary Ethnography, 47*(4), 399–425. https://doi.org/10.1177/0891241617696808

McGivney, V. (2002). *Informal learning in the community: A trigger for change and development* (3rd ed.). National Institute of Adult Continuing Education.

Négrier, E. (2015). Festivalisation: Patterns and limits. In C. Newbold, C. Maughan, J. Jordan, & F. Bianchini (Eds.), *Focus on festivals: Contemporary European case studies and perspectives.* Goodfellow Publishers.

Paris, S. G. (2002). *Perspectives on object-centered learning in museums.* Routledge.

Pimlico Toy Library. (2014). *Our story.* http://www.pimlicotoylibrary.org/our-story/

Rogers, A. (2014). *The base of the iceberg: Informal learning and its impact on formal and non-formal learning.* Budrich.

Rossetti, G., & Quinn, B. (2019). Learning at literary festivals in literary tourism. In I. Jenkins & K. A. Lund (Eds.), *Literary tourism: Theories, practice and case studies* (pp. 93–111). CABI.

Sisson, A. D., & Alcorn, M. R. (2022). How was your music festival experience? Impacts on loyalty, word-of-mouth, and sustainability behaviors. *Event Management, 26.* https://doi.org/10.3727/152599521x16288665119495

SouthWestFest. (2020). *SouthWestFest 16 years strong.* https://southwestfest.org.uk/wp-content/uploads/2020/12/16-YEARS-STRONG-REPORT-Final-PDF_compressed-1.pdf

SouthWestFest. (2022). *About us.* https://southwestfest.org.uk/

Stevenson, N. (2020). The contribution of community events to social sustainability in local neighbourhoods. *Journal of Sustainable Tourism, 29*(11–12), 1–16. https://doi.org/10.1080/09669582.2020.1808664

Veal, A. (2017). *Research methods for leisure and tourism.* Pearson UK.

Viljoen, A., Kruger, M., & Saayman, M. (2018). The art of tastings: Enhancing the arts festival experience. *International Journal of Event and Festival Management, 3,* 246–265. https://doi.org/10.1108/ijefm-03-2018-0023

Westminster Abbey. (n.d.). *Learning.* Westminster Abbey. https://www.westminster-abbey.org/learning

Yoon, Y.-S., Lee, J.-S., & Lee, C.-K. (2010). Measuring festival quality and value affecting visitors' satisfaction and loyalty using a structural approach. *International Journal of Hospitality Management, 29*(2), 335–342. https://doi.org/10.1016/j.ijhm.2009.10.002

Part II
Audience Engagement

6 Music Festivals and Edutainment in a Digital Age

Xiao Lu

Introduction

The 4th China-UK International Music Festival (CUIMF) was live streamed on YouTube and Bilibili in May 2021 due to the COVID-19 outbreak. Since 2018, the festival organisers, including Music at Queen Mary University of London, the London Confucius Institute of SOAS, and the Chinese Classical Instruments Studio, have delivered one of the annual music festivals across nations. The CUIMF is not only a cultural event spotlighting pop music genres and stars, but it features the music of China, Chinese instruments, Western instruments, and hybrid music performances for cultural exchange and music education. The previous events were held in physical venues to present live music performances by Chinese and UK-based bands and musicians. However, the ongoing pandemic remains challenging for live performances and onsite participation. In response to the pandemic situation, the CUIMF has adopted an online practice to design and deliver an international music festival via digital platforms. The use of digital platforms enables the continuity of the CUIMF to organise a yearly event to maintain active engagement with audiences.

With the rapid development of the internet and communication technologies, digital platforms play a vital role in shaping the relationships between audiences and music festivals in today's music and event sectors (Hesmondhalgh & Meier, 2018; Katz, 2010). Digital platforms allow globe-wide audiences to connect with professional artists and music companies, making particular music genres, instruments, and performance styles. In the digital age, both audiences and musicians create the meaning of music and the experiences of music festivals, and in particular audiences become more active and professional to present their voices and opinions by using digital spaces and mobile tools (Ritzer & Jurgenson, 2010). With the global culture flow, rapid technological innovation enhances the quality, diversity, and audience-centred experiences of music for musicians and audiences across borders (Hesmondhalgh, 2021). A range of musical genres and instruments are integrated to create entertaining and educational experiences in international music festivals, which is referred to as edutainment, a coined term for the mixing of entertainment and education (Druin & Solomon, 1996). The CUIMF employs the notion of edutainment to

DOI: 10.4324/9781003305415-8

design the festival by combining music learning and entertainment with the practice of interaction and communication throughout the festival (Aksakal, 2015). This chapter aims to explore the role of digital platforms in creating edutainment by taking the example of the China-UK International Music Festival. Given the situation of COVID-19, the 4th CUIMF needs to deal with the current challenge by employing digital platforms to keep engaging with international audiences.

Through online delivery, the CUIMF aims to build a professional platform to promote cultural diversity and music education between the two countries and beyond. Besides the music programmes, the festival organises academic exchange programmes, including world-leading experts' lectures, masterclasses, and music competitions as ways of inviting audiences to both have fun and learn (Shulman & Bowen, 2001). As the spokesperson of the National Youth Jazz Orchestra explains in the video interview, "For people in China who would like to access the National Youth Jazz Orchestra they can do that through the virtual academy" (Bilibili, 2021). The festival attempts to create an accessible space for audiences to gain entertaining and learning experiences around the world. Digital platforms make it possible to reach local and international audiences who may or may not familiarise themselves with either Chinese or Western music genres, instruments, and musicians, or both. Therefore, this chapter explores how an international music festival uses digital platforms to create edutainment experiences for audiences with diverse cultural backgrounds.

Edutainment and digital festival experiences

Music festivals are one-of-its-kind popular events with the commercial potential to attract fans, music lovers, and fun seekers alike in the world (Linden & Linden, 2018). They tend to present their programmes by mixing a variety of musicians, genres, instruments, performance styles, and company affiliations (Carniel, 2017). In recent years, digital platforms have been applied to introduce cultural events and edutainment content; however, the literature has not to date investigated digital platforms as an increasingly popular venue for mediating unfamiliar cultural and entertainment experiences between audiences and festivals.

Festivals deliver educational value by developing audiences' new cultural knowledge and skills, advancing the appreciation of art and intercultural competence (Delamere, 2001; Ferdinand & Kitchin, 2012). With the expansion of higher education, people who have obtained educational and cultural capital from the university are more likely to pursue cultural products and services to enrich their life experiences and demonstrate their cultural competency and social identity among a group of people (Matheson, 2008). Cultural capital is one of the key factors that shape people's leisure perceptions and preferences reflecting on the choice of cultural goods, services, and lifestyles (Bourdieu, 1984b, 1986). The possession of specialised knowledge that accrues from education, international travel, and professional experience can support people's appreciation of different cultural and entertainment forms. In particular, difficult, high-minded, and overseas cultural forms are appreciated by those who can make sense of them and enjoy them in everyday socio-cultural practices (DiMaggio, 1982). In the age of Web 2.0, the possession of educational and cultural capital also reflects the understanding and mastery of digital tools. Cultural capital is mobilised to transform individuals' digital capital, which shows the ability to access cultural events and festivals via digital and media technologies. Tapscott, Ticoll, and Lowy (2000) claim the development of digital capital allows individuals to exercise their educational and cultural capital on the internet and digital platforms for cultural participation. Those participants have proper skills, confidence, and awareness necessary to access available resources and tools for online experiences and activities (McGillivray & Mahon, 2021). Therefore, educational and cultural capital can be translated to consume different forms of culture and entertainment activities in online and offline settings. Online festival experience is an example that requires audiences to know music techniques and narratives for appreciating the beauty of the performance on digital platforms or physical venues.

Following the expansion of a consumer market, digital platforms are utilised for productive purposes, including entertainment, social networking, consumption, and learning (McGillivray & Mahon, 2021). According to Bourdieu (1984a), cultural intermediaries most typically produce cultural content on TV and radio or act as critics of quality newspapers and magazines. In Bourdieu's view, media plays an indirect cultural mechanism for the reproduction of the social order, and thus the role of cultural intermediaries needs to employ a particular medium platform to accomplish their pedagogic work of cultivating cultural tastes for public engagement and mass consumption (Jones et al., 2019; Matthews, 2010). As Bilton (2017) suggests, the concept of digital intermediaries is introduced in the 21st century, referring to search engines, social media services, broadband providers, and online retailers, which commodify digital resources to build new value chains and business models. Digital intermediaries allow cultural producers to focus on audience experiences and online engagement, which aims to add long-term value to online experiences for potential and existing audiences (Lu, 2020). For today's cultural events, the emergence of digital intermediaries underlines how digital innovation reshapes the process of taste-making and the perception of amusing and educational experiences.

Digital platforms expand the capacity of cultural intermediaries to influence their audiences, and they have become the initial point of contact with new audiences (Negus, 2002). Apart from traditional cultural intermediaries such as television and magazine, digital cultural intermediaries have the ability to introduce music learning and entertainment experiences and connect different actors and stakeholders, such as music producers, audiences, and media professionals. Digital platforms providing video-sharing and streaming functions serve as key digital cultural intermediaries to shape the process of production, distribution, and consumption in transnational music practices. Digital platforms for international music festivals act as gatekeepers to introduce particular music genres, artists, and companies across borders. The adoption of edutainment enables music festivals to direct and influence a broader audience to exercise their music knowledge and learn new musical experiences from an online environment (Anikina & Yakimenko, 2015). Digital music festivals create a more accessible way for audiences to gain edutainment experiences and become an interesting learning process to facilitate active participation and knowledge exchange (Morant & Magalhães, 2021).

China-UK International Music Festival – A case study

The content of the CUIMF has been a free source of music learning via entertainment and online participation, which is different from formal music education (Karlsen, 2009). With the wider accessibility of the internet, digital platforms such as YouTube and Bilibili act as digital cultural intermediaries to redefine the role of the CUIMF in music entertainment and education, which influences audiences' cultural tastes, listening experiences, and consumption preferences (Flynn, 2017). The design and delivery of the CUIMF are to make sure that overseas music genres, instruments, and bands can be heard and appreciated through accessible channels in the digital age (Fairchild, 2014; Ferdinand & Kitchin, 2017). The CUIMF uses digital platforms to provide a broadly pedagogic function by integrating various event activities for a large number of audiences. The CUIMF acts as a vital role in guiding audiences to encounter and understand particular artistic and learning experiences of music (Hutchinson, 2017). More importantly, the CUIMF gives new opportunities to enable the process of online interactions and meaning co-creation more than simply providing products or services (Nowak, 2016). Multiple actors including festival organisers, musicians, audiences, and music companies are linked to participating in online activities for co-creating the festival content and experience together.

In 2021, the CUIMF programmes were premiered each weekend through YouTube and Bilibili between 1 May and 30 May with both platforms acting as the CUIMF's official accounts for the festival's online delivery and public engagement. Both digital platforms allowed audiences to post, view, and add comments and were even more engaged with other participants by using the interactive functions of emojis, likes, dislikes, and sharing. The month-long online festival stimulated the production of user-generated content and in particular for post-festival audiences to revisit the programmes and continue to interact with the online content via digital platforms.

Methodology

To understand and interpret the uses of digital platforms to create edutainment content and experiences for festival audiences, this case study employed netnography to capture data relating to the CUIMF's audience experience. The research was conducted on public platforms based on the observation and analysis of the festival's textual and visual content drawing on the materials of the official accounts on video platforms, the festival websites, streamed video content, recorded interviews, online comments, and participant interactions. Participants were anonymous to ensure the protection of their real and virtual identities throughout the research. Ethical approval was obtained from the researcher's university ethics committee to ensure the research was conducted ethically in accordance with the "Internet Research: Ethical Guidelines 3.0 from the Association of Internet Researchers".

General information was collected about the CUIMF, such as programmes, music genres, the number of views, likes and dislikes, shares, and audience comments, of which a total of 65 online comments (34 comments from YouTube and 31 comments from Bilibili) were pseudonymously collected and analysed for understanding how both Chinese and overseas audiences respond to the edutainment experiences created by the music festival. The collected online comments were categorised by themes that reflected the ways of using digital platforms for edutainment in the music festival. The observation of both digital platforms was intended to understand how audiences use digital platforms to engage with the music festival in internet environments. Throughout the research, the textual and visual materials of the music festival comprising ten music programmes were reviewed and analysed between 17 January and 26 February 2022. The Chinese comments on Bilibili were reviewed and translated by the researcher, who is an English–Chinese bilingual speaker.

Results and discussion

The results of the study found the music festival provides quality audio-visual content for audiences to appreciate particular music genres in an online setting. Based on the observation of ten music programmes, they can be categorised into three types (see Table 6.1), which give an overview of how the music

Table 6.1 Types of music performances in the 4th China-UK International Music Festival

1. Chinese musicians playing Chinese traditional instruments	• *Journey along the Silk Road* • CCOM Plucked Instruments Ensemble Concert
2. Chinese musicians playing Western instruments	• Xuefei Yang Guitar Concert • Journey of violinist Hu Kun: From China to the World: the extraordinary musical • Mingyuan Ruan accordion concert
3. UK musicians or bands playing Western instruments	• Piano and Violin Duo by Dinara Klinton and Michael Foyle • Viviane Plekhotkine • The Yehudi Menuhin School Orchestra Concert • London Philharmonic Orchestra • UK's National Youth Jazz Orchestra Concert (NYJO)

programmes are organised, executed, and promoted on digital platforms. The three types of music performances include Chinese musicians playing Chinese traditional instruments, Chinese musicians playing Western instruments, and UK musicians or bands playing Western music. The festival music programmes show that these devised performances, more or less, indicate the nature of genre integration and crossovers, such as UK musicians and bands playing Chinese music. The UK's National Youth Jazz Orchestra played three Chinese popular folk melodies and turned them into a big band piece with a combination style. The Chinese musicians from the CCOM Plucked Instruments Ensemble Concert performed *The Phantom of the Opera* to show how Chinese instruments present a Western musical classic. The UK's Yehudi Menuhin School orchestra students arranged and played the two pieces of Chinese popular folk music called *Colourful Clouds Chasing the Moon* and *Horse Race*. These examples inform that the CUIMF has made a unique musical journey through cultural exchange and mutual learning between artists and musicians. The diversity of music performances could appeal to broad audience preferences aligned with the ways of music education through online festival participation.

Based on the data collected from YouTube and Bilibili, the festival has reached thousands of audience views by using digital platforms after the launch of the CUIMF. Digital platforms have removed geographic and sociocultural barriers to festival participation to reach a wider audience across borders. Some musicians and their performances have been paid more attention to by audiences on digital platforms as they have received a higher view rate, more comments, and other types of online interactions. Xuefei Yang's guitar concert is an example that has obtained more than 10,000 views on YouTube and 1,666 views on Bilibili by 26 February 2022 (see Table 6.2).

The findings demonstrate that online audiences have expressed their enjoyment and created emotional responses to particular performances and musicians. Audiences left comments on different music performances, including: "Hu Kun's *Violinist from China to the World* is an extraordinary musical journey. I am so excited!" (Participant 1), and "His last piece is cool and moving" (Participant 2). "The music of the Chinese instrument Zheng starts a new chapter. What an audio-visual feast. Feel the love of children and women in the Tang dynasty, and is a story of the changing Tang dynasty. Bravo!" (Participant 3). "The performers (in the CCOM Plucked Instruments Ensemble Concert) are charming. It is a masterpiece of performance" (Participant 4). "It is so refreshing to hear professional musicians offering their views on China-UK music collaboration, to hear their experiences in their words, and to see the incomparable London Philharmonic Orchestra doing what they do best! What a treat" (Participant 4). The CUIMF's musicians act as an instructor of particular music, guiding audiences to understand their performances in a deep and instrumental way. Digital platforms create a closer opportunity for audiences to learn about musicians' experiences behind their performances.

Table 6.2 Audience interactions on YouTube and Bilibili

Platforms	YouTube				BILIBILI			
Interactions Programmes	*Views*	*Likes*	*Dislikes*	*Comments*	*Views*	*Likes*	*Shares*	*Comments*
Journey Along the Silk Road	111	12	0	0	282	14	15	3
NYJO	438	22	0	1	111	3	2	0
Xuefei YANG	11,749	325	6	20	1,666	52	9	14
Michael Foyle & Dinara Klinton Violin & Piano Duet	566	20	0	4	233	6	3	0
Accordion Concert	628	28	0	2	1,017	20	7	11
Hu Kun Violinist from China to the World	213	19	0	3	197	4	4	1
CCOM Plucked Instruments Ensemble Concert	448	25	2	2	457	13	13	2
London Philharmonic Orchestra	835	27	0	1	N/A	N/A	N/A	N/A
Viviane Plekhotkine	223	8	0	1	95	4	0	0
The Yehudi Menuhin School Orchestra Concert	312	9	0	0	270	9	8	0
Total				34				31

The expression of fondness for particular musicians is based on audiences' existing knowledge, which can enhance their learning experiences from particular music performances during the festival. For example, "[Xuefei] is a great human being and an amazing musician. Thanks for the magnificent recital" (Participant 5), and "My favourite classical guitarist – always delivers, never disappoints" (Participant 6). The CUIMF has presented recorded interviews with musicians, which connect online audiences' emotions with musicians in an effective way. Another example is Michael Foyle and Dinara Klinton's Violin & Piano Duet, as one of the audiences wrote, "Great way of improving this Saturday afternoon. Know Dinara's work well but now find Michael Foyle interesting as well" (Participant 7), and "This is so beautiful to listen to, thank you both. I loved your interview too, particularly the part about how you met your violin – can't believe she was just sitting under the piano" (Participant 8). Festival participation has become an important way of increasing audiences' cultural and social capital (Linden & Linden, 2017; Wilks, 2009). From audience comments, it illustrates how they have learned new things from enjoyable music performances by exercising their cultural capital to enhance digital music festival experiences.

Additionally, based on some audience comments, the online delivery of the CUIMF tends to have the advantage of developing audiences from online to offline settings. As one of the comments wrote: "Hope to have the opportunity to see the live performance [of *Journey along the Silk Road*], which will be a visual feast waiting for you". Online performances introduce new experiences to audiences who may not encounter unfamiliar melodies and musicians in any physical setting, but the online delivery of edutainment experience is likely to bring online audiences to onsite participation. Another comment indicates the possibility of attending live performances after this online experience, saying,

> [CCOM Plucked Instruments Ensemble Concert] is an amazing concert. I didn't know how much I miss traditional Chinese music with the modern privilege to hear professionals playing at this place. A joy from start to finish – thank you so much for posting ☺).

Based on the above comments, the organisation of the CUIMF acts as a cultural intermediary via video-sharing platforms where online audiences tend to extend their edutainment experience from an online to onsite setting in the future.

Results have also revealed that audience learning experiences are realised through online interactions during the online festival. Online music festivals can be seen as a potential music educational project that provides festival-related learning experiences to a wider audience. Audiences may not have had an opportunity to appreciate Chinese and Western sounding before, however, they have learnt new musical styles, genres, instruments, and performance techniques through the festival. Musicians play an instrumental role in equipping audiences to understand music instruments, genres, and styles. Leading musicians and artists provide the historical and cultural background of particular instruments and music pieces before their formal performances in the video, as part of the music festival experience. Audiences have more active opportunities to choose the programmes they want to experience on digital platforms. Audience comments appear to show their professional knowledge in appreciating a particular programme, as was a comment:

> Some musicians of Xuefei's level of skill and musicianship have attitudes of utmost superiority and disdain for those people who are less accomplished than themselves ... not just in classical, but also rock, jazz, and pop... This young lady is a master, who is a real person, down-to-earth, observant, insightful musically and willing to share the things which she has learned.... We need more musicians/players/people like Xuefei.
>
> (Participant 9)

Another comment wrote: "I always feel that her classical guitar performance is absorbing the essence of the world, and every time there is the innovation of classical playing skills. That's great" (Participant 10). Xuefei Yang Guitar Concert has received positive comments saying, "Chinese themes on the classical guitar do sound well" (Participant 12) and "I have been listening and in a very humble way attempting to play simple classical guitar's pieces, I love the world of classic guitar music and you (Xuefei Yang)'ve opened up for me. Blessings for you" (Participant 13). However, some music performances were given critical feedback, which is also valuable to examine how online audiences perceive the music experience through the festival. The feedback is of great importance to improve the programme from an audience's perspective in the future by offering better online experiences. As comments wrote: "The accordion performance is a bit exaggerated but the music is good" and "A wonderful performance. There seems to be a problem with the recording quality, and the bottom noise is relatively large" (Participant 11).

Above all, the festival participants mobilise their cultural and digital capital to gain emotional resonance, including the fondness and enjoyment of a particular musician or a music genre, which in turn stimulates the interest and motivation of audiences to learn the skills of appreciating music. As Bourdieu's (1984a, 1984b) account suggests, not only artists and musicians are involved in making cultural products and experiences but also festivals and online participants collaboratively make content and experiences via digital platforms. Festival participants obtain cultural capital as a way of adding value to intellectual development and making something interesting together with others. Music festivals introduce new experiences to audiences who tend to practise active participation to learn unfamiliar musical experiences that may be inaccessible in their daily life or difficult to understand (Aksakal, 2015). The online delivery of the CUIMF creates a professional and academic arena to satisfy audiences' needs for self-growth and home amusement.

The CUIMF is an example that cultivates participants' cultural competency and personal interest to appreciate music performances in a cross-cultural context. The educational aim of the festival is achieved through a variety of entertaining content, including audio-visual materials, official websites, and social media accounts. Audience enjoyment of musicians' performances is a key factor connecting audiences' musical and learning experiences with the festival, and especially some professional comments reflect that audiences' previous music knowledge is fundamental to enhancing their experience of appreciating music performances. Audience enjoyment of the festival further helps transfer a music encounter to a loyal audience, which could be attributed to the practice of edutainment and the efforts of music education the musicians have put in to be a significant resource of the CUIMF for the sustainable development of music festivals.

The analysis of audience comments and online interactions shows that digital platforms are an effective channel to build social and cultural connections between artists, audiences, and cultural organisations for music edutainment experiences. The CUIMF offers a mixture of entertainment and learning content through digital platforms accessible to a wider range of audience members to acquire music-related knowledge and experience, and advance audiences' understanding through the festival and online interactions. The practice of edutainment is to reach cross-border audiences and develop their interests in learning and appreciating Chinese or UK-based musical performances during the festival. Broadly speaking, the use of digital platforms tends to mitigate the impact of the COVID-19 outbreak on a global scale, which indicates a future transformation of the event and music sectors in response to the changes and uncertainties. The CUIMF's practice of edutainment delivers the educational, cultural, and social value of a cultural event with the experiences of change in artistic styles, localities, and technologies in comparison to traditional onsite music festivals. As online comments reflect the experience of edutainment, audiences' enjoyment and fondness of particular musicians and performances on digital platforms are likely to improve audiences' interest to deepen their knowledge of Chinese and Western instruments and music genres. In particular, audiences apply their existing knowledge and skills to appreciate music performances during the festival, and they have also learned new musical styles, genres, instruments, and performance techniques introduced by musicians. However, audience comments also reflect on the use of digital platforms that pose challenges including the lack of quality productions and live experiences, which could raise the awareness of festival organisers to optimise the online experience in the future delivery.

The findings have continued to examine the role of digital platforms as digital cultural intermediaries that proselytise for new tastes and experiential needs by presenting online festivals to a wider audience (Nixon & du Gay, 2002). The CUIMF has created and delivered edutainment experiences through YouTube and Bilibili showing a proper connection between the purpose of edutainment and festival participation. The CUIMF is a free online event that encourages active participation from audiences across the world. Audience experiences of the CUIMF are associated with the exchange of emotions and feelings between different participants. Some critical and professional comments are linked to their edutainment experiences gained from the festival, and in particular, the emotional experiences can drive audiences to pursue new learning experiences during and after the festival. Digital platforms have deeper accessibility of interactions, which not only creates an open environment to offer music learning experiences but also invites all participants to learn from each other through a variety of virtual interactions. The CUIMF through YouTube and Bilibili motivates audiences to contribute to the content, symbolic meanings, and communities that are important to the use of digital platforms in audience engagement.

Conclusion

Digital platforms have empowered contemporary festivals that offer open-eye opportunities for audiences to develop dialogues and interactions with participants across borders. Following Bourdieu (1984a)'s treatment of cultural intermediaries, the role of online festivals acts as digital cultural intermediaries in developing audiences' new cultural tastes and preferences through active participation, online conversations, and real-time interactions, and creating new music consumption patterns and digital experiences. Each music programme is independent to act as a taster for audiences who have little knowledge and experience of a particular music genre, and meanwhile, cultivates audiences' interests and prerequisite knowledge for future onsite attendance (Roberts & Townsend, 2015). Compared to onsite festivals, the online delivery provides a unique environment where audiences could learn how to appreciate world-class music performances and unfamiliar genres while having fun and emotional exchange.

Digital platforms make a borderless space to invite a wider audience to extend their artistic and learning experience during and after festivals when international travel and physical interaction are interrupted by COVID-19 (Jenkins et al., 2009). The use of digital platforms has more advantages to invite cross-border audiences to exchange entertainment and learning experiences with other participants, advancing a new experience through online social interaction. Even though the outbreak of COVID-19 poses a series of challenges and difficulties in the music and event industries, it accelerates a digital transformation of music production, circulation, and consumption and allows us to rethink the role of cultural events and festivals in mediating a relationship between musicians and audiences around the world (Asztalos, 2021). In the face of rapid technological and market changes, the 21st-century music business practice appears to create edutainment value for their audiences, and in particular, for new musical experiences, it needs proper learning opportunities for audiences to foster cultural knowledge and skills to deepen their experiences.

84   *Xiao Lu*

References

Aksakal, N. (2015). Theoretical view to the approach of the edutainment. *Procedia – Social and Behavioural Sciences, 186*, 1232–1239. https://doi.org/10.1016/j.sbspro. 2015.04.081

Anikina, O. V., & Yakimenko, E. V. (2015). Edutainment as a modern technology of education. *Procedia – Social and Behavioural Sciences, 166*, 475–479. https://doi. org/10.1016/j.sbspro.2014.12.558

Asztalos, B. (2021). Digital transformation trends in music teaching and training systems. *Studia UBB Musica, 66*(2), 39–47.

Bilibili. (2021). *UK's National Youth Jazz Orchestra – The 10th session of the 4th China-UK International Music Festival.* https://www.bilibili.com/video/BV1mQ4y1R7yb/? spm_id_from=333.337.search-card.all.click&vd_source=7e3f32deaa3e0bbdf25d1a fd2c627b30

Bilton, C. (2017). *The disappearing product: Marketing and markets in the creative industries.* Edward Elgar Publishing.

Bourdieu, P. (1984a). *Distinction: A social critique of the judgement of taste.* Harvard University Press.

Bourdieu, P. (1984b). *A social critique of the judgement of taste.* Routledge.

Bourdieu, P. (1986). The forms of capital. In J. Richardson (Ed.), *Handbook of theory and research for the sociology of education.* Greenwood.

Carniel, J. (2017). Welcome to Eurostralia: The strategic diversity of Australia at the Eurovision Song Contest. *Continuum: Journal of Media and Cultural Studies, 31*(1), 13–23. https://doi.org/10.1080/10304312.2016.1262089

Delamere, T. A. (2001). Development of a scale to measure resident attitudes toward the social impacts of community festivals, part II: Verification of the scale. *Event Management, 7*, 25–38. https://doi.org/10.3727/152599501108751452

Dimaggio, P. (1982). Cultural entrepreneurship in nineteenth-century Boston: the creation of an organizational base for high culture in America. *Media, Culture & Society, 4*(1), 33–50. https://doi.org/10.1177/016344378200400104

Druin, A., & Solomon, C. (1996). *Designing multimedia environments for children: Computers, creativity and kids.* John Wiley and Sons.

Fairchild, C. (2014). Popular music. In J. M. Maguire & J. Matthews (Ed.), *The cultural intermediaries reader.* SAGE Publications.125–133.

Ferdinand, N., & Kitchin, P. (2012). *Events management: An international approach.* SAGE Publications.

Ferdinand, N. and Kitchin, P. J. (2017). *Events management: An international approach* (2nd ed.). Los Angeles: Sage.

Flynn, M. (2017). Accounting for genre: How genre awareness and affinity affects music streaming use. In H. Barlow & D. Rowland (Eds.), *Listening to music: people, practices and experiences.* Open University.

Hesmondhalgh, D. (2021). Streaming's effects on music culture: Old anxieties and new simplifications. *Cultural Sociology*, 1–22. https://doi.org/10.1177/17499755211019974

Hesmondhalgh, D., & Meier, L. (2018). What the digitalisation of music tells us about capitalism, culture and the power of the information technology sector. *Information, Communication & Society, 21*(11), 1555–1570. https://doi.org/10.1080/1369118X.2017. 1340498

Hutchinson, J. (2017). *Cultural intermediaries: Audience participation in media organisations.* Palgrave Macmillan.

Jenkins, H., Rurushotma, R., Weigel, M., Clinton, K., & Robison, A.J. (2009). *Confronting the challenges of participatory culture: Media education for the 21st century*. MIT Press.

Jones, P., Perry, B., & Long, P. (2019). *Cultural intermediaries connecting communities: Revisiting approaches to cultural engagement*. Policy Press.

Karlsen, S. (2009). Learning through music festivals. *International Journal of Community Music, 2*(2–3), 129–141. https://doi.org/10.1386/ijcm.2.2-3.129

Katz, M. (2010). *Capturing sound: How technology has changed music*. University of California Press.

Linden, H., & Linden, S. (2017). *Fans and fan cultures: Tourism, consumerism and social media*. Palgrave Macmillan.

Linden, H., & Linden, S. (2018). "There were only friendly people and love in the air": Fans, tourism and the Eurovision Song Contest. In C. Lundberg & V. Ziakas (Eds.), *Handbook of popular culture and tourism*. Routledge.

Lu, X. (2020). The disappearing product: Marketing and markets in the creative industries. *Cultural Trends, 29*(3), 256–258. https://doi.org/10.1080/09548963.2020.1799333

Matheson, C. M. (2008). Music, emotion and authenticity: A study of Celtic music festival consumers. *Journal of Tourism and Cultural Change, 6*(1), 57–74. https://doi.org/10.1080/14766820802140448

Matthews, J. (2010). Cultural intermediaries and the media. *Sociology Compass, 4*(7), 405–416. https://doi.org/10.1111/j.1751-9020.2010.00285

McGillivray, D., & Mahon, J. (2021). Distributed digital capital: Digital literacies and everyday media practices. *Media Practice and Education, 22*(3). 196–210. http://doi.org/10.1080/25741136.2021.1899628

Morant, F., & Magalhães, L. (2021). Edutainment on YouTube: Tom Scott Channel. In *ECSM 2021 8th European conference on social media*. Academic Conferences Inter.

Negus, K. (2002). The work of cultural intermediaries and the enduring distance between production and consumption. *Cultural Studies, 16*(4), 501–515. https://doi.org/10.1080/09502380210139089

Nixon, S., & du Gay, P. (2002). Who needs cultural intermediaries? *Cultural Studies, 16(*4), 495–500. https://doi.org/10.1080/09502380210139070

Nowak, R. (2016). *Consuming music in the digital age. Technologies, roles and everyday life*. Palgrave MacMillan.

Ritzer, G., & Jurgenson, N. (2010). Production, consumption, prosumption: The nature of capitalism in the age of the digital "prosumer". *Journal of Consumer Culture, 10*, 13–36. https://doi.org/10.1177/1469540509935467

Roberts, E., & Townsend, L. (2015). The contribution of the creative economy to the resilience of rural communities: Exploring cultural and digital capital. *Sociologia Ruralis, 56*(2), 197–219. https://doi.org/10.1111/soru.12075

Shulman, J. L., & Bowen, W. G. (2001). *The game of life: College sports and educational values*. Princeton University Press.

Tapscott, D., Ticoll, D., & Lowy, A. (2000). *Digital capital: Harnessing the power of business webs*. Harvard Business Review Press.

Wilks, L. J. (2009). *Initiations, interactions, cognoscenti: Social and cultural capital in the music festival experience* [Doctoral dissertation, The Open University].

7 Science in Society

Exploring Science Festivals and Valuable Leisure

Elspeth A. Frew and Amaia Makua

Introduction

Science festivals were conceived to try to bring science closer to society (Von Roten, 2011), with their main objectives being to raise public awareness of science, to promote the dialogue between science and society, and to encourage young people to select science as a career (EUSCEA, 2005). Science festivals are different from other science communication activities, such as lectures, public debates, or open doors at academic institutions, because they celebrate science in an informal and festive atmosphere (Bultitude et al., 2011). They also apply novel methods to give non-specialists a chance to glimpse the "inside story" of what it is like to experience science (Fikus, 2007) and they focus on working with topics in a popular manner, and this has helped to attract large audiences to these events (Nolin et al., 2003).

Similar to cultural festivals, science festivals are increasingly including activities related to the performing and visual arts (Durant, 2013). Such artistic mediation involves people emotionally and can facilitate the understanding of science and promote participation and dialogue processes (Von Roten & Moeschler, 2007). In turn, this helps make these public science events more "meaningful, entertaining, and instructive" (Durant, 2013, p. 2681). The range of activities provided are classified as hands-on activities, lectures/talks, discussion/dialogue, and plays/concerts (Bultitude et al., 2011). For example, the Edinburgh Science Festival has been very innovative in using food, drink, music, and gaming to engage different audiences (Gage, 2001), with the festival brochures revealing the variety and extent of the activities. (Edinburgh Science Foundation, 2022). This diversity of content is related to the multidisciplinary character of modern science festivals, and the use of edutainment (reflecting a convergence of education and entertainment) (Komarac, Ozretic-Dosen, & Skare, 2020) in a leisure setting can appeal to visitors (Packer & Ballantyne, 2004). The assortment of activities and the promotion of local interests play a main role in the success of these events. As Durant (2013, p. 2681) states, "Such festivals don't need to be big and expensive; what they need to be is creative and relevant ... endlessly and wonderfully diverse".

DOI: 10.4324/9781003305415-9

Kim et al. (2013) highlight that events provide not only tangible outcomes such as money but also intangible benefits created by positive visitor experiences. In this context, festivals have become universally important for their social and cultural roles, and increasingly they have been promoted and created as tourist attractions (Dwyer & Wickens, 2011) and have useful economic benefits (O'Sullivan & Jackson, 2002). However, up until now, no attempt has been made to relate science and science festivals to another major issue of postmodern society, namely the leisure phenomenon and the associated use of edutainment to appeal to attendees. This chapter reviews valuable leisure theory (Cuenca, 2003, 2014a, 2014b), which reflects nearly 30 years of research and formative work at the Institute of Leisure Studies, University of Deusto, Bilbao, Spain. This chapter then focuses on the British Science Festival, organised by the British Science Association, which develops science engagement programmes to ensure that science is more relevant, representative, and connected to society (British Science Association, 2022) and in so doing this illustrates that a leisure setting such as a science festival has the potential to offer educational components with the communication messages enhanced through edutainment (Packer & Ballantyne, 2004). Thus, science festivals have the potential to engage with attendees with no pre-existing interest in science, and the chapter demonstrates that science festivals are occasions where education and entertainment can occur simultaneously.

Valuable leisure

Leisure is an important concept in everyday life and is becoming recognised as a subjective, valuable, and profound experience over a lifetime (San Salvador del Valle, 2000). Various theories consider the sense of leisure in changing societies (see, e.g. Edginton, 2007) and some theories, like the experience economy, focus on experiences as having market value (Pine & Gilmore, 1999). Others like the humanist leisure theory (Cuenca, 2003) and the valuable leisure theory (Cuenca, 2014b) develop a more comprehensive approach, which suggests that the leisure experience is a complex, personal, and social issue.

Humanist leisure occurs when satisfying and pleasurable experiences are intimately linked to one's system of values and meanings (Cuenca, 2003). It is based on the three fundamental values of autotelic leisure, namely freedom, satisfaction, and gratuity (having an end in itself). It also involves identity, self-improvement, and justice. Valuable leisure further develops the human development aspect whereby leisure experiences have "positive values for the people and for communities, a leisure based on the recognition of the importance of the successful experiences and their potential for social development" (Cuenca, 2014b, p. 87).

Valuable leisure may reflect one or more of the following human needs: satisfaction, enjoyment, separation from reality, or for personal self-development. The valuable adjective emphasises the beneficial social value that is recognised in the practice of certain leisure activities, as well as their potential for human development, but which does not exclude other types of benefits, such as economic development (Cuenca, 2014b). As Cuenca (2014b, p. 87) points out, valuable leisure is "multi-faceted and diverse, depending on the perception of practitioners and values and potentialities that, objectively, develop". Thus, the humanist and valuable leisure theories are aligned with the idea of improving not only subjective but also collective wellbeing and happiness. Leisure contributes to making those collective goals a reality (Cuenca, 2014a).

The theory of valuable leisure emphasises the contribution of the leisure experience to human development through the following: enhancement of an individual's capabilities, the promotion of their wellbeing, the empowerment of the individual, the encouragement of conviviality among participants, the support of an individual's values, and the protection of a region's cultural diversity. Alongside these ideas of happiness and wellbeing, the theory of valuable leisure highlights the importance of the political, economic, social, cultural, and educational circumstances, which facilitate, hinder, or limit developing valuable leisure experiences (Instituto de Estudios de Ocio, 2013) (see Table 7.1).

Table 7.1 Key elements of the valuable leisure experience

Key Elements	Characteristics
Development of capabilities	Leisure allows an ongoing process for the identification, enhancement, and improvement of an individual's capabilities
Protection of cultural diversity	Leisure is part of the rich and complex cultural heritage of communities
Promotion of wellbeing	Leisure can develop sensitivity and perception, increase creativity, and promote physical and emotional health, thus prolonging life
Promotion of empowerment	Leisure can empower individuals through positive and satisfying experiences that are freely chosen. Leisure can empower communities through the promotion of creative and supportive leisure that promotes self-improvement, participation, economic development, identities, and citizens' wellbeing
Ensuring conviviality	Leisure supports conviviality; promotes interpersonal relationships inspired by equity, solidarity, and inclusion; and allows creative and communication skills as well as critical attitudes among citizens
Promotion of values	Leisure promotes freedom, satisfaction, and intrinsic motivation, interrelated with identity, achievement, and justice. Freedom here refers to the intrinsic freedom that facilitates choice and, particularly, to the actual leisure freedoms enjoyed by individuals and communities

Adapted from Instituto de Estudios de Ocio (2013).

The challenge of communicating science to society

Science communication is designed to enable the general public and science practitioners to interact with each other more effectively and aims to help develop "an interest in science, a confidence to talk about it, and a willingness to engage with science wherever and whenever it crosses their paths" (Burns, O'Connor, & Stocklmayer, 2003, p. 89). Thus, science communication has a vital role in modern society because a gap between scientists and community could result in scientific ignorance; may help create a breeding ground for "alienation, demagogy and extremism" (Bauer, 2009, p. 223); and may cause citizens to reject new advances (Fikus, 2007). Furthermore, "science cannot develop if society is unable to grasp its objectives, methodology, and output" (Fikus, 2007, p. 48). Science events and science festivals are becoming central to informal science communication worldwide (Neresini, Dimopoulos, Kallfass, & Peter, 2009) as they bridge science and society in a "non-intimidating, festive framework, thus being able to reach a wider audience" (Von Roten, 2011, p. 6), and science festivals are becoming one of the most prevalent informal science engagement events at an international level (EUSCEA, 2005).

In recognition of the importance of events and festivals as experience drivers and the potential role of science in society, the objective of the present research is to determine the extent to which a science festival such as the British Science Festival uses a variety of techniques such as edutainment where education and education converge (Rapeepisam et al., 2006) to promote the conditions that enable participants to experience a transforming, educational, emotional, memorable leisure experience that is entertaining, educational, and fosters their personal and their community's development. To address this objective, the following questions were posed based on the key concepts of the valuable leisure theory (Instituto de Estudios de Ocio, 2013):

> Can a science festival use various elements of edutainment to 1) allow individuals to develop capabilities? 2) Help protect cultural diversity? 3) Promote wellbeing? 4) Promote individual and community empowerment? 5) Ensure conviviality? 6) Promote values?

The British Science Festival – A case study

The British Science Festival is organised by the British Science Association and is described as Europe's longest-standing science festival as it was first held in York in 1831. The festival is held annually in cities across the United Kingdom and hosted by various universities. For example, in 2022, it was hosted by De Montfort University, Leicester, and in 2023 by the University of Exeter. The British Science Festival is designed to connect the general public with scientists, engineers, technologists, and social scientists and is designed to engender discussion and debate, and given the rise in the number of science festivals nationwide it appears to have inspired many other science festivals in the UK to be established (British Science Festival, 2022).

Methodology

A descriptive case study approach was used (Yin, 2013) and a documentary method allowed the gathering of the data via participant observation at the site of the 2013 British Science Festival held in Newcastle and via public domain sources (including newspapers, web pages, and government reports) (Hodder, 1994), in this case via the British Science Festival programme and webpage accessed online. The case study approach allowed a close analysis of the secondary data associated with the science festival and allowed the researchers to consider both the manifest and latent content of specific texts. Themes were generated by comparison between data and inferences were made. The festival was examined within the framework of the valuable leisure theory (Cuenca, 2003) to determine the extent to which the festival represented a valuable leisure experience. The data was then manually coded and subjected to thematic analysis.

The public festival programme contained a detailed description of the 183 activities held during the British Science Festival, and this became the major source of data for the analysis (Newcastle Science Festival, 2013). The organisers of the British Science Festival provided a range of categories for each of the activities: by day, by audience type, by type of event, and by area of science. The present research utilised the categorisation of audience type and type of event as this reflected the target market and the activities designed for these audience members. These sources were analysed using various forms of textual analysis and content analysis, a research technique for "making replicable and valid inferences from texts (or other meaningful matter) to their context of their use" (Krippendorff, 2003, p. 18). Textual analysis and content analysis are useful methodologies for deconstructing the content of festivals and have recently been utilised at other festivals (see e.g. Whitford & Ruhanen, 2013). Despite using the 2013 programme for analysis, a review of the most recent festival in 2021 revealed a great deal of similarity in terms of content and activities.

The researchers analysed the content of the festival programme (see https://www.newcastlesciencecity.com/british-science-festival/) and examined the language and expressions used to engage the audience). Each of the events was examined in relation to the following: the event title; the formal or informal nature of the phrasing; the sentence structure; and the verbs, adjectives, and adverbs used. As a result, each activity held during the festival was analysed in terms of its ability to engage, excite, and entertain the audience within the framework of the valuable leisure theory. Critical and recurrent issues were identified and further investigated using the framework of the valuable leisure experience. In accordance with these findings, inferences were made regarding evidence of the facilitation of valuable leisure theory via the British Science Festival. The following discussion is divided into two main sections, namely the activities included in the programme and the descriptors used.

Results

Activities included in the programme

The most common activities at the British Science Festival were "Talks and Debates" (39%) and "Family Events" (28%). An example of a Talk or Debate is "When Fridges Attack: Big Data Meets Intelligent Machines", which discussed the first recorded case of a smart device, a fridge, being hacked and used to send spam emails and the speakers reflected on how technology improve our lives but recognised that it opens a host of new dangers. An example of a Family Event was "Science of Star Wars", which looked at the science of droids and spaceships and how scientists can make real lightsabers. This activity included a Star Wars character quiz and a prize for the best-dressed youngster. The other festival activities occurred less frequently namely: "Evening Events" (10%), "Exhibitions & Drop-In" (9%), "Trips and Tours" (8%), and "Adult Workshops" (5%).

Because the target market for the festival was wide-ranging (i.e. academics, general public, school groups, and families of all interests and backgrounds), the festival activities were designed to be attractive to a broad and non-specialist audience, mainly composed of adults. Activities for families represented 18% of the programme (Families with Teenagers, 11%, and Families with Children, 7%). The remaining target audiences represented 2% and 1% of the activities (Adults with Some Knowledge and Adults with In-depth Knowledge). This focus reinforces the aim of the festival to reach out to the general public as the main audience of their activities were non-specialists, namely All Adults.

The Talks and Debates activities allowed the attendees to become aware of the diverse issues associated with science and technology and were closely connected with day-to-day issues. For example, there was a lecture on the impact of lack of exercise on preventable diseases like diabetes and obesity, and another example focused on converting food waste into fuel. Therefore, the audience could feel some identification with these topics. The Talks and Debates ranged from maths to politics, and from religion to history and nutrition. This activity in the programme had less physical involvement by the attendees, but the few that were creative involved very engaging activities that combined science and the arts. For example, there was a comedy show called "Festival of the Spoken Nerd", which involved three stand-up comedians. The use of humour as a form of play is a positive aspect of human communication and creates a mode of interaction that can lead to amity (Sciama, 2016), helping to make the event more engaging and entertaining.

Family Events for Everyone were varied, and they allowed attendees to become involved in different ways and depth. Apart from fostering the popularisation of science topics, they also tried to promote the identification or the development of attendees' capabilities (Instituto de Estudios de Ocio, 2013) through challenging activities. One example of these engaging activities was "Jesmond Dene Bioblitz", in which the participants had to help the organisers find and identify species of insects in 24 hours. These Family Events for Everyone demonstrated a creative and socialising spirit. Some activities were very imaginative, such as "Explore the Evolutionary History of Teddy Bears", which invited attendees to bring their teddy along to a teddy bears' picnic to explore their history. These sorts of creative activities are multidisciplinary as they combined science with other cultural and leisure disciplines, such as music, sport, and theatre. This contributed to raising the social appreciation of science as well as to promoting local culture and cultural diversity. Furthermore, these activities tried to get closer to the citizens by literally taking science to the streets. There were several activities that took place in the streets and shopping centres that tried to draw the public's attention by making science a spectacle. For example, "Street Science", "Newcastle University Street Scientists", and "Ingenious Busking" allowed trained science buskers to impress the crowds with their close-up science demonstrations in public venues, such as streets, shopping centres, and the university campus. As such, conviviality and accessibility are patent in these events.

Family Events for Families with Teenagers provided a range of challenging activities. These activities attempted to involve the attendees in events such as "The Poison Project". Some of these activities were designed and communicated in a very creative way. Some others relied on popular speakers from the media, who may have had a greater capacity to engage these teenagers. Family Events for Families with Children aimed to be enjoyable and creative. Such activities fostered the development of perception and capabilities; for example, "Great thinkers and do-er's treasure hunt". Some of them relied on spectacular views of science and on the current popularity of these scientists among younger citizens; for example, "The Improbable Zoo".

Evening activities

The evening activities mainly comprised activities exclusively for adults. However, there were some evening events organised for families with teenagers, reflecting the benefit of science festivals to use their popularity and accessibility via edutainment to reach those hard-to-access demographics. Evening Activities for All Adults principally aimed at developing an understanding and appreciation of diverse science issues. Some of these evening events dealt with health issues or helped the audience appreciate science. The most participative examples were also the most creative ones, for example, "Anatomy Life Drawing", "Brain Festival (Pub Quiz)", and "Electromagnetic Pulse Party". In these examples, science interfaced with other cultural disciplines, such as arts, live music, and comedy.

Exhibitions and drop-in

The exhibition and drop-in events were generally open to everybody and tried to popularise science and to engage a broad audience of non-specialists with science. These activities frequently combine science with other cultural interests, such as music, comics, and art, for example, "The Sonic Cosmos: A Music Exploration into Our Galactic Neighbourhood". Most of these activities involved creativity. Furthermore, these events merged science with other cultural expressions to reinforce the social and cultural appreciation of science. Although these were static exhibitions, they were attended by scientists who answered visitors' questions. These exhibitions allowed visitors to drop in and spend some non-curated time wandering around these exhibits. Phrases like "drop into the basement" highlighted the impromptu nature of these visits.

Trips and tours

Trips and tours were designed for the general and family public. There were two categories: All Adults and Families and Teenagers. In the festival, the Trips and Tours provided a range of opportunities for visitors to tour various sites of importance in the local area. During these tours, opportunities were provided to engage with the scientists and to see first-hand the physical evidence of science in the community. Some of these tours used the expression "Behind the Scenes" tours. These trips and tours ranged from visiting an observatory to visiting Roman remains and experiencing nature first-hand. These tours allowed visitors the opportunity to visit the environment and to experience first-hand science artefacts in the outdoors. These provided a link with the local environment and reinforced that science exists in a physical sense. The trips for All Adults involved the participants in a special way and promoted interpersonal relationships (conviviality). The trips for Families with Teenagers and Children aimed to involve attendees in a different way, making them aware of interesting issues through the visits to technological facilities. This helped to foster awareness and personal relationships.

Adult workshops

These adult workshops mainly tried to engage non-specialist audiences (All Adults). However, there were a few activities that tried to engage people with a more advanced level of science knowledge. These adult workshops were pitched at a higher level than those for teenagers. For example, some involved sampling aspects, for example, tasting an octopus curry and handling real fossils. The workshops designed for All Adults involved participatory activities that tried to popularise science. These workshops generally dealt with topics that may affect everyday life of the attendees such as health or wellbeing topics, for example, "drugs and sport", and "How macrobiotic is your yoghurt?" The workshops designed specifically for Adults with Some Knowledge promoted perception, and participation and were concerned with day-to-day matters such as finance. The only workshop for Adults with In-depth Knowledge provided a teamwork event designed to encourage new ideas. These new ideas could be developed due to the deep involvement of the varied attendees, who may have ranged from businesspeople to researchers and investors. This activity could have an impact on the economic development of the region and may reflect community empowerment.

Descriptors used

The organisers of the British Science Festival tried to attract the target audience using direct, friendly, and informal language and engaging expressions. Throughout the programme, the language used in the descriptors ensured the content was very accessible to a broad audience, and readers had an increased chance of understanding the content as references were made to a range of problems and issues experienced in everyday life. In addition, the adjectives, verbs, and pronouns used ensured the topics sounded very interesting and engaging. The titles and the descriptors also used an array of colloquial expressions and layman terms to enhance understanding by everyone. This reflects the open nature of the festival, designed to allow everyone to celebrate science regardless of his or her background. Examples of this informal use of language are "Learn weird and wacky facts about the human body"; "Who hasn't heard of Google?"; and "What's hiding in your mouth?"

The language used by the organisers is inclusive and creates ownership. On the one hand, the programme frequently used a range of personal pronouns (namely, "we", "us", "you") and possessive adjectives (namely, "your", "our"). The use of these types of words helps to engage the reader more directly: for example, "Space scientists … take *you* on a Journey of Discovery"; or "*We* can investigate it". In addition, it allows people to feel involved and to value their opinion, despite their non-expertise in science. Other examples of such expressions are reflected in the following examples: "What would *you* do?"; "What do *you* think?"; "How many can *you* find and complete?".

There were also numerous examples of the use of imperative verbs, which give instructions and, in this case, encouraged the audience to participate in the festival. These imperative verbs appeared in the titles and also in the descriptors of the activities. For example, "*Save* our fish stocks" and "*Make* your own homunculus". Some examples of imperative descriptors are: "*Find out* how Itch uses his brains", "*Come and listen* to experts", "*Learn* some basic facts", and "*Join* our team of scientists". Thus, the festival programme contained various engagement, educational, and entertainment techniques used by the festival agencies to encourage the public to participate in the science festival. To sum up the activities and descriptors included in the British Science Festival, the programme showed evidence of innovative and creative events which pushed some boundaries, tried to engage with the attendees physically and emotionally, and included content that was relevant and applicable to the attendees' everyday life. Thus, there was evidence of valuable leisure activities at the science festival created using edutainment techniques and elements.

Discussion

The following discussion considers the extent to which the British Science Festival provided a valuable leisure experience and the use of edutainment to allow individuals to develop capabilities, help protect cultural diversity, promote wellbeing, promote individual and community empowerment, ensure conviviality, and promote values (Instituto de Estudios de Ocio, 2013).

To allow individuals to develop capabilities

The British Science Festival used a variety of educational techniques to allow individuals to develop capabilities (Instituto de Estudios de Ocio, 2013) through the encouragement of individuals to engage in learning activities. In so doing, the individuals had the opportunity to learn something new that was often related to problems in the world around them and the scientific solutions, which could be used to overcome them. These sorts of activities are challenging and frequently demand an intellectual effort by attendees. Through such engagement, the attendees could identify, enhance, and improve their capabilities particularly as they were "forced" to move out of their comfort area, to participate in these activities, and to engage with the festival more fully. Such personal undertaking is one of the premises to have a valuable leisure experience under the theory of Humanist Leisure (Cuenca, 2014b). The more cultivated, the better the person is prepared to understand and encourage the activity, the more satisfaction he/she receives and the more meaningful that experience becomes.

Protection of cultural diversity

The science festival demonstrated protection of local identity (Instituto de Estudios de Ocio, 2013) by introducing educational and engaging activities that highlighted the local environment and the history and culture of the region. Furthermore, the festival programme includes other countries' cultural expressions and gives the target audiences exposure to them. These activities provided the opportunity for the local community and the visitors to travel into the local region to experience and appreciate first-hand the cultural diversity which exists around them. In addition to this, a range of creative activities were developed to encourage a greater relationship between science and other cultural expressions, such as the performing arts (music, theatre, comedy), fine arts (painting), literature (comics), and popular media (TV and newspapers). In so doing, this highlights that science can be a cultural expression, which can reinforce its importance and relevance to society. In addition, some of the innovative activities raise the appreciation for science as a socio-cultural expression. This appreciation is the first step to rebuild public trust in science, as stated in the "Science-in-Society" concept.

Promotion of wellbeing

The existence of the British Science Festival (and other similar festivals around the world) uses a variety of educational techniques to provide the general public with an increased awareness of the relationship between science and the usefulness of science to encourage good living, longevity, and wellbeing (Instituto de Estudios de Ocio, 2013). There were many examples of events and activities in the programme that were educational, informative, and insightful about health-related issues. The festival clearly aimed to develop sensitivity and perception of science through activities related to the day-to-day life of attendees. The content of numerous engaging activities was aimed at promoting both physical and emotional health. Furthermore, the creative focus of the programmed activities invited the participants to become involved and develop their own creative skills.

Promotion of individual and community empowerment

The mere existence of the British Science Festival recognises the importance of community and individual empowerment (Instituto de Estudios de Ocio, 2013). This event offers a wide range of creative and educational activities for different audiences. Having so many options to choose from means that the target audience needs can be met and the innovative activities are designed to provide fulfilling experiences. Many of these activities are challenging to the audience mentally, physically, or both, with the traditional views of the audience members being challenged. For example, such challenges were identified in relation to the evidence of climate change, alternative energy supplies, human mortality, space and time, genetics, and mental diseases, all of which are relevant to the general public and their everyday lives.

Getting the maximum participation both physically and emotionally are two of the main priorities of these kinds of festivals. The language used to communicate the programme used a direct and appealing style. Apart from the language, the festival offered a wide range of hands-on activities that demanded an active role by the attendees, reflecting various elements of edutainment. In addition, community empowerment was encouraged as the festival was organised by a consortium of local and national agents and some activities tried to generate new ideas by engaging with local stakeholders such as the local university, the city council, the city library, local museums, a local cinema, and a pub. These ideas have the potential to form the basis for future socio-economic projects in the region.

Ensuring conviviality

The festival spirit of this event encourages and supports conviviality (Instituto de Estudios de Ocio, 2013) and interpersonal relationships among attendees.

The celebration of science takes place through inclusive activities for all types of members of the public whereby they have the chance to engage with like-minded individuals, to perhaps work as a team, and to create a feeling of comradeship among participants. These educational activities also encourage important critical thinking as they invite people to question prejudices related to science that may affect their lives.

Promotion of values

Science has great potential to encourage free living and satisfaction with intrinsic experiences. The educational activities provided at the science festival can also help establish strong connections with local culture and strengthen the identity of citizens within a region. Furthermore, these activities and experiences have the power to challenge individuals and societies to think of additional ways of developing themselves, reflecting that the science festival supported the promotion of values (Instituto de Estudios de Ocio, 2013).

Conclusion

The use of various aspects of edutainment and learning through play, such as the use of technology, visual materials, and indoor and outdoor settings (Komarac et al., 2020; Packer & Ballantyne (2004), reflects the pivotal role of science festivals in contemporary society. Similar to Packer and Ballantyne (2004), the science festival appears to have created an educational experience which was entertaining and exciting. This chapter demonstrates that science festivals, such as the British Science Festival, contribute to raising popular awareness about science and are the starting points for bridging science and citizens' leisure. Indeed, other festivals and events could use science festivals as an exemplary way to reach out to hard-to-reach demographics in an entertaining and educational manner. In this context, access to science can be a source of satisfactory experiences, which can contribute to personal and collective development, as well as to greater levels of quality of life and happiness. This chapter uses the example of the British Science Festival held in Newcastle, United Kingdom, to explore the extent to which a valuable leisure experience was created and the extent to which the science festival promoted the conditions that enable participants to experience a transforming, emotional, memorable leisure experience that had the potential to foster their personal and their community's development. An examination of the festival programme revealed that the activities and the spirit with which they were raised, demonstrated an approach aligned with the key aspects of valuable leisure experience. Those activities facilitated the promotion of values, the development of capabilities, the protection of cultural diversity, the promotion of wellbeing, the promotion of empowerment, and, conviviality (Instituto de Estudios de Ocio, 2013). Furthermore, the festival organisers used a range of techniques to engage the audience members, including providing a range of creative and innovative activities and using language to encourage audience members to become involved.

Considering the humanist and valuable theories (Cuenca, 2003, 2014a, 2014b), the provision of these activities at the festival provided opportunities to develop meaningful, memorable, and transforming experiences (Pine & Gilmore, 1999). Consequently, stakeholders involved in the festivals and events management sector should bear in mind this quest and facilitate the conditions to develop these sorts of experiences by attendees. Even though this event was designed to make science accessible to all citizens, the programme was heavily skewed towards adults, reflecting the overall purpose of the science festival to reach the general public. However, the science community could expand the programme to include younger people and families to encourage them to independently explore science without being directed by schools. Future research in this domain could compare the philosophy and types of activities programmed at other national festivals, such as the national science festivals which already exist in the USA, Holland, and Japan, and explore the expectations of attendees before the science festival and their reflections after their visit. This would help festival organisers more fully create events to meet the needs of the attendees and ensure that science is portrayed as exciting, interesting, and relevant in today's society.

References

Bauer, M. (2009). The evolution of public understanding of science – Discourse and comparative evidence. *Science, Technology & Society, 14*(2), 221–240. https://doi.org/10.1177/097172180901400202

British Science Festival. (2022). *Celebrating the people, stories and ideas at the heart of science.* https://britishsciencefestival.org/

Bultitude, K., McDonald, D., & Custead S. (2011). The rise and rise of science festivals: An international review of organised events to celebrate science. *International Journal of Science Education, Part B, 1*(2), 165–188. https://doi.org/10.1080/21548455.2011.588851

Burns, T. W., O'Connor, D. J., & Stocklmayer, S. M. (2003). Science communication: A contemporary definition. *Public Understanding of Science, 12*, 183–202. https://doi.org/10.1177/09636625030122004

Cuenca, M. (2003). *Ocio humanista. Documentos de Estudios de Ocio, 16.* Universidad de Deusto.

Cuenca, M. (2014a). Aproximación al ocio valioso. *Revista Brasileira de Estudos do Lazer. Belo Horizonte, 1*(1), 21–41.

Cuenca, M. (2014b). *Ocio Valioso. Documentos de Estudios de Ocio, 52.* Universidad de Deusto.

Durant, J., (2013). The role of science festivals. *Proceedings of the National Academy of Sciences, 110*(8), 2681–2681.

Dwyer, L., & Wickens, E. (2011). Event tourism and cultural tourism: issues & debates: An introduction. *Journal of Hospitality Marketing & Management, 20*(3/4), 239–245. https://doi.org/10.1080/19368623.2011.562408

Edginton, C. R. (2007). Leisure as social transformation. *Journal of Leisure Studies*, *5*(1), 1–12.

Edinburgh Science Foundation. (2022). *Brochure downloads*. https://www.sciencefestival.co.uk/brochure-downloads

EUSCEA. (2005). *White book on science communication events in Europe*. European Science Events Association.

Fikus, M. (2007). A festival against ignorance. *ACADEMIA: The Magazine of the Polish Academy of Sciences*, *3*(15), 48–49.

Gage, S. (2001). Edinburgh International Science Festival. In S. M. Stocklmayer, M. M. Core, & C. Bryant (Eds.), *Science communication in theory and practice* (pp. 203–217). Springer.

Hodder, I. (1994) The interpretation of documents and material culture. In N. K. Denzin and Y. S. Lincoln (Eds.) *Handbook of qualitative research* (pp. 393–402). Sage.

Instituto de Estudios de Ocio. (2013). *Manifiesto por un ocio valioso para el Desarrollo Humano*. http://www.asociacionotium.org/wp-content/uploads/2014/03/Manifiesto-por-un-Ocio-Valioso-para-el-Desarrollo-Humano.pdf

Kim, J., Boo, S., & Kim, Y. (2013). Patterns and trends in event tourism study topics over 30 years. *International Journal of Event and Festival Management*, *4*(1), 66–83. https://doi.org/10.1108/17582951311307520

Komarac, T., Ozretic-Dosen, D., & Skare, V. (2020). Managing edutainment and perceived authenticity of museum visitor experience: Insights from qualitative study. *Museum Management and Curatorship*, *35*(2), 160–181. https://doi.org/10.1080/09647775.2019.1630850

Krippendorff, K. (2003), *Content analysis: An introduction to its methodology* (2nd ed.). Sage Publications.

Neresini, F., Dimopoulos, K., Kallfass, M., & Peter, H. (2009). Exploring a black box: Cross-national study of visit effects on visitors to large physics research centers in Europe. *Science Communication*, *30*, 506–533. https://doi.org/10.1177/1075547009332650

Newcastle Science Festival. (2013). *British Science Festival*. https://www.newcastlesciencecity.com/british-science-festival/

Nolin, J., Bragesjö, F., & Kasperowski, D. (2003). Science festivals and weeks as spaces for OPUS. In E. Felt (Ed.), *O.P.U.S: Optimising public understanding of science and technology—Final report* (pp. 271–282). University of Vienna.

O'Sullivan, D., & Jackson, M. J. (2002). Festival tourism: A contributor to sustainable local economic development? *Journal of Sustainable Tourism*, *10*(4), 325–342. https://doi.org/10.1080/09669580208667171

Packer, J., & Ballantyne, R. (2004). Is educational leisure a contradiction in terms? Exploring the synergy of education and entertainment. *Annals of Leisure Research*, *7*(1), 54–71. https://doi.org/10.1080/11745398.2004.10600939

Pine, B. J., & Gilmore, J. H. (1999). *The experience economy: Work is a theatre and everyday business a stage*. Harvard Business School Press.

Rapeepisarn, K., Wong, K. W., Fung, C. C., & Depickere, A. (2006). Similarities and differences between "learn through play" and "edutainment". In *Proceedings of the 3rd Australasian Conference on Interactive Entertainment*, 4–6 December 2006, Perth, WA, pp. 28–32.

San Salvador del Valle, R. (2000). *Políticas de ocio*. Universidad de Deusto.

Sciama, L. D. (Ed.). (2016). *Humour, comedy and laughter: Obscenities, paradoxes, insights and the renewal of life (Vol. 8 of social identities)*. Berghahn Books.

Von Roten, F. C. (2011). In search of a new public for scientific exhibitions or festivals: The track of casual visitors. *Journal of Science Communication, 10*(1). https://doi.org/10.22323/2.10010202

Von Roten, F. C., & Moeschler, O. (2007). Is art a 'good' mediator in a science festival? *Journal of Science Communication, 6*(03), A2. https://doi.org/10.22323/2.06030202

Whitford, M., & Ruhanen, L. (2013). Indigenous festivals and community development: A sociocultural analysis of an Australian indigenous festival. *Event Management, 17*(1), 49–61. https://doi.org/10.3727/152599513X13623342048149

Yin, R. K. (2013). *Case study research: Design and methods.* Sage.

8 Maintaining Festivals' "Sense of Community" through Educational Activities for Locals and Festival Tourists during the COVID-19 Crisis in Greece

Sofoklis Skoultsos and Nicholas Karachalis

Introduction

Festivals are inherent to the sociability of humans, holding a particular role in enforcing a bond between their attendees. At the same time, there is also a connection between festivals and local development goals, tourism, and sustainability and often a regenerative role acclaimed to events, which allows for discourse concerning the impact assessment of festivals (O'Connor, 2015; Wilson et al., 2020). The supply of festivals has grown over the past few decades, which has led to the growth of a strong professionalisation of festival management, audience development, and festival marketing (Noelle Bernick & Boo, 2013). Festivals are increasingly connected to non-cultural, economic benefits, development goals, and regeneration programmes within the visitor economy. Educational and participatory characteristics are often part of commercial, hybrid festivals, such as Burning Man in Nevada, USA – a festival that combines economic performance with spectacle, personal engagement, and wellness. Festival attendees' motives and social goals often go beyond entertainment: community bonding, self-improvement, and social cohesion become main motivations for many (Getz, 2012; Getz & Page, 2016; Lawendowski & Besta, 2020; Rowen, 2020). Studies have shown that smaller festivals are mostly connected to social capital and the idea of combining a festival with participation, education, and sociability (Bakas et al., 2019; Mair & Weber, 2019; Stevenson, 2016; Yolal et al., 2012). These efforts can lead to a connection with edutainment, which Wyatt (2022) describes as a way for visitors to experience heritage sites, events, and destinations, to change from being spectators into participants, further arguing that these experiences are often translated into higher consumer value and increased revenue for the organisers. Still, as discussed in this chapter, smaller festivals that include edutainment are mostly linked to a sense of belonging and socialisation, which often becomes the main motive of attendee participation.

Drawing on this, this chapter looks into the way community festival organisers in Greece combined cultural production and consumption with educational functions. It further relies on the organisers' points of view to explore

DOI: 10.4324/9781003305415-10

how both the organisers and the audience evaluate the interaction and the sense of belonging created in the context of a community festival. Based on field research by the authors, this chapter presents key evidence that the educational role of community festivals can contribute in an effort to confront the effects of COVID-19 and allow the festival organisers and participants to interact and maintain sociability.

Community festivals and sense of community

While the effects of the COVID-19 pandemic have been a major shock for most forms of social and economic activity, leisure activities, culture, and tourism have been impacted in a disproportionate and more severe way (Banks, 2020). This is quite evident with festivals in particular, as they are reliant on people coming together to attend or celebrate a spectacle or performance. COVID-19 occurred during a time when major transitions, such as digital and green transitions, were occurring within the festival world and affecting the way live performances and festivals operate. Yet, it is still difficult to estimate the long-term effect that the pandemic will have on everyday life, including sociability and the way people interact during festivals. The inevitable closing down of festivals due to the pandemic has allowed for an opportunity to test the resilience of the festival management structures, as well as to assess the loyalty of festival audiences and their need to participate in festivals that promote an edutainment experience. This has been particularly evident for smaller festivals and community festivals.

Festivals have always offered a way to socialise and feel part of a community, which has become a strong element for many festivals in smaller cities in Greece (Konsola & Karachalis, 2010). The main objectives of community festivals include creating conditions to involve members of the community in an artistic experience, linking the festival to local characteristics, celebrating local partnerships and promoting local distinctiveness, and raising the place's profile regionally and internationally (Rossetti & Quinn, 2019, 2021). Apart from these characteristics, community festivals also need to adjust to multiple factors that affect the current sector conditions, which are discussed in the following.

Community festivals are highly affected by the sense of community or sense of belonging (Szabó, 2015). Putnam (2000) describes how social capital is associated with civic, social, and political life, and points out two basic elements of social capital: bonding and bridging. Bonding and bridging can be considered as key elements for community festivals as well (Wilks, 2011). Temporality is also seen as a key element of festivals, which means that the gained social capital does not lead to a permanent sense of community (Carvalho, 2020; Comunian, 2017). Supporting this, the digital transition, which is partly enforced by the pandemic, has led to hybrid festivals and the use of online media to help enhance the interaction between participants outside the festival period (Seraphin, 2021).

Another essential factor is connected to the general tendency to lead management structures of cultural institutions to become more participatory within a crowdsourcing approach. This also concerns a turn towards involving different stakeholders' representatives in the management of festivals. On this topic, Wallace and Michopoulou (2018) present an extended list of stakeholders (from sponsors and local decision makers to hotel owners) and evaluate their involvement, influence, and interest concerning events and festivals.

Resilience is also key for festivals because while insecurity and inflexible working conditions were the norm for many years, COVID-19 highlighted these difficult conditions and the need for resilience through new support measures on local or national levels (Oakley & Ward, 2018). One of the key questions stemming from this, which is addressed in this chapter, is whether smaller community festivals suffered more from the pandemic crisis effects. Finally, it is widely agreed that the visitor economy is shifting towards a model where visitors are no longer seen as passive participants, but instead seek out co-creation experiences (Wyatt, 2022) thus extending the need for more edutainment experiences within community festivals.

Edutainment and festivals

The educational part of cultural and creative industries in general, and festivals in particular, is a strong element for both their popularity and economic success. Recent publications on festivals show that film, music, and performing arts festivals are increasingly aiming to inspire learning (Biaett & Richards, 2020; Mair & Duffy, 2018; Ziakas, 2016). In wine festivals, for example, researchers have argued that in terms of knowledge and skills, festival attendees should be engaged through active participation in activities relating to the tasting and wine-making process (Fernandes & Cruz, 2016; Saayman & van der Merwe, 2015). The notion of edutainment (educational entertainment) is central to this discussion as well as theming, which is a central element to edutainment, given the active and participatory experience. Specifically, edutainment refers to interaction, increased enthusiasm, and excitement of the learner, applying knowledge and enjoying learning (Aksakal, 2015). Mair and Duffy (2018) also refer to the entertainment nature of edutainment, specifically in creating the groundwork for networks and allowing a sense of belonging and relationships, leading to a condition where social capital becomes the central element.

Greek festivals and community – A case study

Literature on festivals in Greece has shown an increase in the number of festivals in the past two decades that rely on theming. These festivals are organised outside the metropolitan areas, mostly in the summer period, and are often connected to tourism. Konsola and Karachalis (2009) argue that for most of these festivals, education and the support of a local cultural scene are key objectives, as opposed to economic or touristic development. However, the link to tourism must not be underestimated, as the majority of festivals are organised in places that also attract tourism during the summer period (Skoultsos, 2014). As such, recent research has highlighted particularities of festival tourists in Greece (Skoultsos et al., 2020). Drawing on this, this chapter explores cases of Greek community festivals that use edutainment as a tool to improve and enhance the sense of community among the attendees during and after the COVID-19 pandemic. In an effort to contribute to the discussion on festival management in Greece during and after the pandemic through the lens of edutainment practices, a research hypothesis is formed: *community festivals in tourism destinations outside the large metropolitan areas showed greater resilience during the pandemic crisis because of their sense of community, their focus on edutainment, and their informal business approach.*

The scope of the research examines how educational activities, through edutainment, supported the sense of community of festivals and increased the viability of the festivals, especially during times of the pandemic crisis. The research focuses on seven community festivals in Greece that have developed a sense of community due to their orientation, strategies, and implementation practices. The following provides relevant information for each of the selected cases.

The Lefkas Music Week – a music festival, financed by tuition fees and local sponsors and run by an NGO. Rooted in the musical tradition of the island, the festival has entered its 16th edition, focusing mostly on classical music. Its educational offerings include masterclasses on violin, viola, cello, singing, piano, conducting, chamber music, and orchestra, alongside concerts in unusual places of the island (e.g. abandoned villages). It partners with the local public music school, Orfeas (Mousikologikos Omilos Orfeas), the municipality, and the local tourism businesses to create a group of stakeholders that support the event and its educational objectives. Its success is bolstered by communication outreach and enhanced visibility through networks and delegations. The festival is classified as a small community festival as it has only two employees.

Primarolia – a cultural community festival organised in Aigio in the Peloponnese and designed around the legacy of the Corinthian currant. Now in its fourth edition, the festival is run by an NGO but is organised by a network of people who live in the area or are connected to it. The festival's intention is to offer a unique experience for attendees through activities that include performances, screenings, tours, and exhibitions – all with a strong educational agenda. Creating partnerships on a local level is a key element for this festival, which allows for an economic value in terms of sponsorship and the provision of spaces.

Marpissa Festival – Paros – a cultural festival that offers experiential activities to its audience. The festival's aim is to create thematic routes in the traditional village of Marpissa at Paros Island (Greece), where attendees play an active role in the festival activities. Managed by a group of 25 volunteers, the festival runs throughout the year. However, during July and August these volunteers increase to approximately 80 in total.

Music Village Ag. Lavrentios is a music festival based on workshops. The main idea of the festival is to establish an annual meeting for music and performing arts focused mainly on the participation of professional or amateur musicians. The festival takes place in a small village on the mountain of Pelion; it was first held in 2007. The event management team consists of musicians and volunteers that support the implementation.

Molyvos International Music Festival (MIMF) – a four-day classical music celebration focused on young talents, initiated eight years ago. It takes place in the unique village setting of Molyvos with the main goal to promote music coexistence of highly acclaimed artists. Managed by a team of 15, the festival offers concerts, organised at different stages throughout the village, and various activities to strengthen the bonds with the local community.

Vovousa Festival – a three-week cultural festival that takes place in the mountainous area of Epirus. Running for eight years, the main goal of the festival is to promote sustainability, nature conservation, and the promotion of the host destination. Run by a small team of volunteers, the festival offers various activities related to video art, photography, workshops for children, text reading, etc., which take place in the village of Vovousa.

Anima Syros Festival – a film festival that began in 2008 and has since shown over 2,700 animated films from all over the world. Run by a team of ten employees, which grows to 30 during the festival's delivery period, the goal is to promote the art of animation through educational workshops, film presentations, and various networking activities by operating as a hub of animation professionals at a pan-European and international levels.

All the selected cases take place in different parts of Greece and represent a wide range of different rural destinations such as islands and mountains. The criteria for the selection of the sample were the following:

- Host destination: Rural area
- Sense of community: Established
- Years of existence: At least three years of successful implementation
- Activities: Including educational activities and workshops

Methodology

In order to gather data, the authors used secondary data through websites and online sources, and conducted semi-structured interviews with the directors and managers of the selected festivals. Interviews were held through online sessions, which varied in duration from 30 to 60 minutes, and focused on the following issues: general information about the festival, strategy and goals, sense of community, educational activities, and impacts of COVID-19.

In order to comply with ethics relating to voluntary participation, confidentiality, and anonymity, all participants provided written consent agreement to participate in the research.

A limitation of the research is that the audience's perspective has not directly been taken into account. To include these perspectives, on-the-spot research with questionnaires would be necessary. Another limitation is connected to the lack of evidence on the role of online communities around festivals. While there are few indications documented, a netnographic approach would be more appropriate for exploring this data, which is suggested as a next step for this research.

Results and discussion

In review of the data analysis, key themes for discussion were identified: the role of edutainment as a strong element to create loyalty among the festival followers, the sense of community as a strength, and how the strong participatory culture can be connected to the resilience of the festivals during the pandemic crisis.

Edutainment and sense of community

The festivals explored for this study developed activities to enhance the audience's participation and support the interaction between them and the local population through workshops and edutainment activities. Through this, they created a strong sense of community and implemented policies that create strong bonds between the audience and the local community.

According to the manager, *Marpissa festival* establishes memorable and holistic experiences by offering active participation in workshops related to local culture and traditions. From the onset, the management team wanted to establish an interaction between the volunteers, locals, and the audience through the festival activities using active engagement and participation in the workshops. Connecting the volunteers and local people with the audience allows interaction and building of social bonds. Active participation is thus used to establish a memorable and holistic experience as the audience can meet locals and other members of the audience. Interaction is thus a significant element of the Marpissa festival. Yet, this engagement is not simply observed during the festival period as volunteers, locals, and local schools gather cultural heritage materials to present to the audience during the festival activities. It is important to note that this engagement and interactivity between the audience and locals is enhanced by the fact that the workshops take place on several islands in the Aegean Sea, thereby allowing locals from all islands to participate and strengthening their social bond with the festival audience. By doing this, the festival is able to limit the commercialisation of the festival and build on the sense of community, mainly through the inclusion of local people and the active participation of the audience – which is a key element of edutainment activities. The organisers explained that these types of activities establish a memorable and holistic experience because the audience can meet local people and the rest of the audience. Interaction is a very important thing as it offers the opportunity to participate as one in the experience.

The manager of the *Vovousa Festival* commented that the festival is based mainly on workshops and on the mixing of educational activities and entertainment (i.e. edutainment), with the main goal being to promote the destination and offer an opportunity for the audience to experience local traditions. In the early stages of the festival, the festival's main focus was the concerts that took place over eight days. However, during the last four years, the festival's activities have become the key focus, with workshops showcasing local traditions (e.g. preparing local food, presenting the local flora of healing herbs) and the destination. It is important to mention that these workshops also operate as a source of income for the management team to cover the expenses of the whole festival. In addition, the Vovousa Festival engages local houses and people to welcome and accommodate visitors, which helps to create a strong sense of belonging as part of the festival, thereby leading to repeat visitation by the audience.

AnimaSyros Festival also offers workshops related to animation, which are largely promoted to local schools. The edutainment concept is implemented through workshops that take place at several outdoor places (e.g. beaches) during the summer period. The festival activities and workshops promote a positive image of the festival to the local population and the audience, with an increasing number of local people supporting the festival as volunteers. Anima Syros Festival is thus based on the sense of community for people interested in animation films. Moreover, the festival is an annual meeting for professionals and amateurs related to animation. This sense of an animation community is further enhanced by the strong bond established with the local community of Syros Island and traditional culture of the wider area of Cyclades.

Primarolia Festival aims to develop a participatory festival experience, which according to the manager-artistic director, is mostly achieved by combining events that are addressed to a wider audience with activities that require an active participation (e.g. participants can commit to an oral history group concerning the particular industrial building the festival is organised in each time). Special attention is dedicated to activities for children and families from the area, allowing a community to develop around the festival. Volunteers also play a key role in community building. According to the organiser, the festival aims to attract a specific niche audience and not the wider public, with the goal of allowing the attendees to get involved, participate, and interact. The manager explains, "We are not interested in attracting a wider crowd, we prefer to have less followers that understand and embrace our character and informal way of work". Also, the festival has developed a network through partner organisations such as the Danish Archaeological Institute of Athens and the Open University.

Lefkas Music Festival has a strong educational focus including master-classes on violin, viola, cello, singing, piano, conducting, chamber music, and orchestra. According to the manager, the educational part ensures financial resilience, and many musicians have their costs managed by their employer. Many of the educational activities are directed at professionals or advanced participants; however, others, such as musical amateurs or beginners, may also try out their musical/vocational skills. Moreover, the outcome of the educational program is presented to the audience, allowing both an interaction between the participants and a direct way of experiencing the music. The Lefkas Music Week is based on the Ionian islands' particular legacy in orchestra and choir music. This legacy creates acceptance and pride within the local community; therefore, the festival is not seen as something which is organised "outside of the community". The thematic focus of the festival, the foreign musicians, and the international recognition of the performers create the circumstances to attract a wider audience.

It is important to mention that all the respondents highlighted the importance of promotion practices throughout the year to contact on a constant basis with the audience and local communities. Online interaction through social media and concerts and activities to local communities and schools are some of the common practices that the selected festivals implement to further build a sense of community.

Interaction and participation as resilience factors

All the respondents declared that the enhanced sense of community supported the resilience of their festival even during COVID-19. The workshops and activities discussed above strengthened the bonds between the festivals' audience and the local communities. This fact resulted in increased loyalty and repeat visitation and at the same time anticipation to meet again after the "pause" due to COVID-19 protection measures. For the manager of the Lefkas Music Festival, "the informal personal relations between participants create trust and allow a more informal relation". The same experience is shared by most of the managers: the people who are responsible for the educational part of the festivals feel part of the local community, combine their work with vacations/relaxation time, and are committed to the goals of the event (Table 8.1). A key outcome of the research is connected to the fact that these festivals do not follow the typical business model of a commercial festival, allowing staff and participating artists to develop a different relationship. Edutainment, as a core element of all the examined festivals, can also be seen as a strong participatory element for all the involved parties (managing team, artists, local stakeholders).

Table 8.1 Perspectives of festival managers of the selected festivals

Title	Sense of Community	Edutainment	Interaction and Participation as Resilience Factors
Lefkas Music Festival	Based on the Ionian islands' particular legacy in orchestra and choir music that create community pride	Masterclasses on violin, viola, cello, singing, piano, conducting, chamber music, and orchestra for adults/children	Enhanced experience Loyalty Repeat visitation Financial resilience
Primarolia Festival	Developed networks through partner organisations such as the Danish Archaeological Institute of Athens, the Open University, and groups such as fine art students, and local schools	Active participation of visitors is promoted through the oral history workshop on the current tradition, participatory mobile filmmaking workshops for families, etc.	Enhanced experience Loyalty Repeat visitation
Marpissa Festival (Diadromes Marpissa)	Local people are included actively in the event management teamAudience can meet local people and the rest of the audience through active participation	Cultural heritage material is gathered throughout the year by volunteers, local people, and local schools. The material is used either for workshops or enhancing the experience of the festival attendees during the summer period	Enhanced experience Loyalty Repeat visitation
Molyvos Music Festival	Development of Association of "Friends of the Festival" to connect fans of the festival	Training activities and educational programs related to cultural management (e.g. art and music-kinetic workshops, The MO-TO Key)[1]	Enhanced experience Loyalty Repeat visitation
Animasyros	Annual meeting for professionals and amateurs related to animation	Animation workshops throughout the year at local schools	Enhanced experience Loyalty Repeat visitation
Music Village	Community of professional and amateur musicians	Workshops of performing arts (mainly music) and concerts	Loyalty Repeat visitation Financial resilience
Vovousa Festival	Engages local houses and people to welcome and accommodate visitors	Workshops that mix educational activities and entertainment	Enhanced experience Loyalty Repeat visitation Financial resilience

1 "The MO-TO key programme was launched in 2017. It aims to introduce children to cultural management. The MO-TO Key is carried out in collaboration with the German cultural institution TONALi and the Regional Directorate of Primary and Secondary Education of the North Aegean".

Conclusions

The relationship between education, tourism, and community festivals, being both strong and controversial, is a very challenging one during the pandemic crisis. Inevitably, community festivals have the sense of community as their core objective and as a result they depend on participation and interaction between local people and the audience. Though the examined festivals differed regarding their initial goals, character and management structure were common elements detected regarding their response to the pandemic. Edutainment, linking workshops and arts, and engaging the audience in participatory activities were common practices that enhanced the resilience of the festivals, especially during the negative impacts of COVID-19. Despite the relevant limitations of the current study, the examined festivals have the following common elements:

- They share a strong educational focus which leads to a dedicated community. Professional or amateur musicians, filmmakers, actors, etc. who have been following the works of each festival soon, even as visitors initially become part of the organisation.
- They depend on partnerships on a local level: funding, provision of services, provision of performance spaces, promotion activities, etc. are all provided by partner institutions.
- Leadership and professional management are highlighted as a key element of the selected festivals. For example, even though community festivals depend on voluntary work in all cases, there is a core group that has a professional relation with the event. The managers also expressed their willingness to further professionalise. In the case of the Molyvos festival, the organisation of cultural management workshops implies an interest to further work on management issues. The informal relations between managers and participating artists, the commitment to the festivals' goals, the low dependence on ticket revenue, and the strong educational focus were, to a great extent, the reason that during the difficult times of the pandemic crisis the festivals didn't collapse and showed resilience.
- On a management level, and especially during the COVID-19 pandemic, they depend on peer-to-peer information and guidance. For example, the director of the Lefkas Music Festival is in contact with the festival organiser of the Music Village of Agios Lavrentios in order to exchange best practices.
- Although developing a tourism profile is not an initial motive, a community festival can become part of a different, more sustainable, and community-based (or even community-owned) form of tourism development, for example, slow tourism, and sustainable tourism. The cases examined in this chapter highlight the importance of the sense of community that is enhanced through edutainment and participatory activities.

The above findings are also in line with findings of recent works on the future of festivals after the COVID-19 pandemic (e.g. Davies 2021; Fotiadis et al., 2021;

Seraphin 2021). The resilience of the festivals is still a debatable issue in terms of the negative impacts of the economic and post-COVID crisis. Nevertheless, practices that strengthen the bonds between local community and visitors such as the ones that are related to edutainment remain an important element that can lead to event and festival viability in the long term (Rossetti & Quinn, 2022). As derived from current studies, experiences that promote interaction and edutainment have a strong positive effect on the attractiveness of each festival. These findings comply with previous research related to edutainment (see e.g. Wyatt, 2022). In this context, edutainment can play a significant role for festival managers regarding increased active participation, which can lead to strengthening bonds between artists, attendees, and local people.

References

Aksakal, N. (2015). Theoretical view to the approach of the edutainment. *Procedia – Social and Behavioral Sciences*, *186*, 1232–1239. https://doi.org/10.1016/j.sbspro.2015.04.081

Bakas, F. E., Duxbury, N., Remoaldo, P. C., & Matos, O. (2019). The social utility of small-scale art festivals with creative tourism in Portugal. *International Journal of Event and Festival Management*, *10*(3), 248–266. https://doi.org/10.1108/IJEFM-02-2019-0009

Banks, M. (2020). The work of culture and C-19. *European Journal of Cultural studies*, *23*, 648–654. https://doi.org/10.1177/1367549420924687

Biaett, V., & Richards, G. (2020). Event experiences: Measurement and meaning. *Journal of Policy Research in Tourism, Leisure and Events*, *12*(3), 277–292. https://doi.org/10.1080/19407963.2020.1820146

Carvalho, R. (2020). Understanding the creative tourism experience in cultural and creative events/festivals. *ISLA Multidisciplinary e-Journal*, *3*(1), 1–18.

Comunian, R. (2017). Temporary clusters and communities of practice in the creative economy: Festivals as temporary knowledge networks. *Space and Culture*, *20*(3), 329–343. https://doi.org/10.1177/1206331216660318

Davies, K. (2021). Festivals post Covid-19. *Leisure Sciences*, *43*(1–2), 184–189, https://doi.org/10.1080/01490400.2020.1774000

Fernandes, T., & Cruz, M. (2016). Dimensions and outcomes of experience quality in tourism: The case of Port wine cellars, *Journal of Retailing and Consumer Services*, *31*, 371–379. https://doi.org/10.1016/j.jretconser.2016.05.002

Fotiadis, A., Polyzos, S., & Huan, T. (2021). The good, the bad and the ugly on COVID-19 tourism recovery. *Annals of Tourism Research*, *87*, 103–117. https://doi.org/10.1016/j.annals.2020.103117

Getz, D. (2012). *Event studies. Theory, research and policy for planned events*. Routledge.

Getz, D., & Page, S. J. (2016). Progress and prospects for event tourism research. *Tourism Management*, *52*, 593–631. https://doi.org/10.1016/j.tourman.2015.03.007

Konsola, D., & Karachalis, N. (2009). Arts festivals and urban cultural policies: The case of medium sized and small cities in Greece. In S. Ada (Ed.) *Cultural policy and management yearbook* (pp. 51–63) Istanbul Bilgi University Press.

Konsola, D., & Karachalis, N. (2010). The creative potential of medium-sized Greek cities: Critical reflections on contemporary cultural strategies. *International Journal of Sustainable Development*, *13*(1–2), 84–96. https://doi.org/10.1504/IJSD.2010.035101

Lawendowski, R., & Besta, T. (2020). Is participation in music festivals a self-expansion opportunity? Identity, self-perception, and the importance of music's functions. *Musicae Scientiae, 24*(2), 206–226. https://doi.org/10.1177/1029864918792593

Mair, J., & Duffy, M. (2018). The role of festivals in strengthening social capital in rural communities. *Event Management, 22*(6), 875–889 https://doi.org/10.3727/1525995 18X15346132863229

Mair, J., & Weber, K. (2019). Event and festival research: A review and research directions. *International Journal of Event and Festival Management, 10*(3), 209–216. https://doi.org/10.1108/IJEFM-10-2019-080

Noelle Bernick, L., & Boo, S. (2013). Festival tourism and the entertainment age: Interdisciplinary thought on an international travel phenomenon. *International Journal of Culture, Tourism and Hospitality Research, 7*(2), 169–174. https://doi.org/10.1108/IJCTHR-04-2013-0023

O'Connor, J. (2015) Intermediaries and imaginaries in the cultural and creative industries. *Regional Studies, 49*, 374–387. https://doi.org/10.1080/00343404.2012.748982

Oakley, K., & Ward, J. (2018). Creative economy, critical perspectives. *Cultural Trends, 27*, 311–312. https://doi.org/10.1080/09548963.2018.1534573

Putnam, R. (2000). *Bowling alone: The collapse and revival of American community.* Simon and Schuster.

Rossetti, G., & Quinn, B. (2019). Learning at literary festivals. In I. Jenkins & L. A. Lund (Eds.), *Literary tourism: Theories, practice and case studies* (pp. 93–105). CABI.

Rossetti, G., & Quinn, B. (2021). Understanding the cultural potential of rural festivals: A conceptual framework of cultural capital development. *Journal of Rural Studies, 86*, 46–53. https://doi.org/10.1016/j.jrurstud.2021.05.009

Rossetti, G., & Quinn, B. (2022). The value of the serious leisure perspective in understanding cultural capital embodiment in festival settings. *The Sociological Review.* https://doi.org/10.1177/00380261221108589

Rowen, I. (2020). The transformational festival as a subversive toolbox for a transformed tourism: Lessons from Burning Man for a COVID-19 world. *Tourism Geographies, 22*(3), 695–702. https://doi.org/10.1080/14616688.2020.1759132

Saayman, M., & van der Merwe, A. (2015). Factors determining visitors' memorable wine-tasting experience at wineries. *Anatolia, 26*(3), 372–383. https://doi.org/10.1080/13032917.2014.968793

Seraphin. (2021). COVID-19: An opportunity to review existing grounded theories in event studies. *Journal of Convention & Event Tourism, 22*(1), 3–35. https://doi.org/10.1080/15470148.2020.1776657

Skoultsos, S. (2014). The potential of festivals and their contribution to culture and tourism. *Greek Economic Outlook, 25*, 61–66.

Skoultsos, S., Georgoula, V., & Temponera, E. (2020). Exploring "Sense of Community" in the festival tourism experience: Review of the relative literature. In V. Katsoni & T. Spyriadis (Eds.), *Cultural and tourism innovation in the digital era. Springer proceedings in business and economics.* Springer.

Stevenson, N. (2016). Local festivals, social capital and sustainable destination development: Experiences in East London. *Journal of Sustainable Tourism, 24*(7), 990–1006. https://doi.org/10.1080/09669582.2015.1128943

Szabó, J. Z. (2015). Festivals, conformity and socialisation. In C. Newbold, C. Maugan, J. Jordan, & B. Franco (Eds.), *Focus on festivals: Contemporary European case studies and perspectives* (pp. 40–52). Goodfellow Publishers Limited.

Wallace, K., & Michopoulou, E. (2018). The stakeholder sandwich – A new stakeholder analysis model for events and festivals. Presentation at *Tourism, Hospitality and Events International Conference (THEINC)*, Buxton, UK.

Wilks, L. (2011). Bridging and bonding: Social capital at music festivals. *Journal of Policy Research in Tourism, Leisure and Events*, *3*(3), 281–297. https://doi.org/10.108 0/19407963.2011.576870

Wilson, N., Gross, J., Dent, T. C., Conot, B., & Comunian, R. (2020). *Re-thinking inclusive and sustainable growth for the creative economy: A literature review*. DISCE Publications.

Wyatt, B. (2022). Edutainment. In D. Buhalis (Ed.), *Encyclopedia of tourism management and marketing*. Edward Elgar Publishing.

Yolal, M., Woo, E., Cetinel, F., & Uysal, M. (2012). Comparative research of motivations across different festival products. *International Journal of Event and Festival Management*, *3*(1), 66–80. https://doi.org/10.1108/17582951211210942

Ziakas, V. (2016). Fostering the social utility of events: An integrative framework for the strategic use of events in community development. *Current Issues in Tourism*, *19*(11), 1136–1157. https://doi.org/10.1080/13683500.2013.849664

Part III
Sustainability and EDI

9 The Winterage Festival in a Learning Landscape

Angela Wright

Introduction

This chapter explores an edutainment experience lived by tourists who often seek more than a cursory visit to a destination. The festival and edutainment experience discussed here focuses on a region in County Clare, Ireland, called The Burren. The Burren (spread across north Clare, south-east Galway, and the Aran Islands) is halfway along the west coast of Ireland and is a landscape like no other. Stark, sheeted, and riddled with cracks on glacial limestone, it is home to an abundance of flora and fauna, fossils, cliffs, and caves. Flowers pigmented in striking shades from deep purple to radiant gold dot the lunar-like limestone canvas, their origins traced back to far-flung lands, such as the Caribbean and Egypt. Due to this incredible landscape, the Burren has gained international fame for its landscape flora and fauna, with many areas designated as special areas of conservation. Not only can visitors enjoy the unusual habitat, but the rich and distinctive historical element provides the final compelling component. In the built environment, over 2,700 recorded monuments are to be found, some dating back over 6,000 years (burren.ie, n.d.). The region includes the Cliffs of Moher, and the area has UNESCO Global Geopark Status.

Among The Burren's many regular documented visitors was the English writer J. R. R. Tolkien (1892–1973), author of *The Hobbit* and *The Lord of the Rings*, who is said to have been inspired by The Burren's singular topographical appearance in later writings. Tourists readily understand the potential for inspiration with local festivals adding another enticement to travel and experience the region and landscape.

Exploring the Burren Winterage festival, this chapter examines how this festival aligns farming practices, nature, and education. Here, co-creation (Wright, 2019) is at its best with tourists, locals, culture, traditions, and food providing an occasion where education and entertainment can take place simultaneously – edutainment. This chapter investigates how the festival evolved as an edutainment experience, as well as how participants, organisers, and the festival audience consume this festival. In this landscape, visitors seek experiences that are informative and educational (Cabanas, 2020), while being captivated by the intensity of nature.

DOI: 10.4324/9781003305415-12

Along with the local community who give their time and venues for free, the weekend is coordinated by stakeholders such as the local landscape charity, The Burrenbeo Trust (independent charity advocating placed-based learning), The Burren Irish Farmers Association, and The Burren Farming for Conservation Programme. Support is also provided by The Heritage Council and the Department of Agriculture, Food and the Marine, and FBD Insurance (burrenwinterage.com, n.d.). Burrenbeo has helped engage the broader Burren community through monthly walks/talks, volunteering events, and festivals, including The Burren Winterage Weekend celebrating the rich legacy of pastoral farming (Dunford, 2016).

Contributing to this piece, Áine Bird, Burrenbeo Trust Coordinator and Researcher, says that The Burren is a "man-made" landscape. It exists because of the influence and interaction of humans over thousands of years, particularly farming, and the local specific practice of Winterage. Winterage is an aged practice where livestock are moved to the uplands during the winter, where they graze back strong vegetation. When they return to the lower pasture in the summer, there is space and light for the growth of The Burren flora and fauna to emerge.

Place, emotions, and space, combined on a similar level, are powerful (Cabanas, 2020). Cabanas (2020), examining space and emotion in the setting of theme parks, calls for new related studies. This case study on the Winterage Festival fits this narrative and provides hues of these concepts with the theme park here being a natural landscape setting – termed "a learning landscape" (Dunford, 2016, p. 36). A clear and defined relationship is found between the space and place of The Burren landscape and the emotions of those that inhabit, nurture, protect, visit, and earn there. This study is novel in that it examines this relationship and edutainment in the context of a small festival in a "learning landscape" (Figure 9.1).

Festivals

A festival is a form of a play and the play is the culture (Kim, 2017, p. 183). Festivals, for Ireland, are not a new concept with many dating back to Celtic and Pagan times. Many of these religious festivals were bound as one with nature and the seasons, such as *Imbolc, Beltane, Lughnasadh*, and *Samhain* (Irishmyths.com, n.d.). Mention of these festivals can be found in old Irish literature dating to the 9th century. Past mass gatherings from political to religious celebrations and gatherings to modern-day music festivals and traditional festivals such as the Saint Patrick's Day Festival (17 March annually) define what it is to be Irish. Festivals are a vital element for the growth of tourism worldwide, and, in an Irish context, festivals attract approximately 200,000 overseas visitors to Ireland contributing €108 million (Fáilte Ireland, 2019) to the Irish economy annually during normal years. Fáilte Ireland recognises that festivals are major contributors to the economic well-being of the Irish tourist offering and an important marketing tool internationally (Wright, 2009, 2017). Hence, Fáilte Ireland has supported and promoted Irish festivals, as they recognise the value of such events. In 2019, the Irish Government

Figure 9.1 The cattle drive.
Source: (Burrenbeo).

provided €2,964,500 in Fáilte Ireland funding to enhance and promote 40 festivals (Fáilte Ireland, 2019) and develop new festivals and events across the country. Kim (2017) outlines that the revitalisation of local festivals which allow the increment of one's most personal happiness index is estimated to have the most crucial significance. Rossetti and Quinn (2021, p. 53) found rural festivals engage with rural community heritage: "Festivals are sources of inspiration, knowledge-sharing, enjoyment, and social networking that feed cultural imaginations and enrich lifestyles in rural places".

The ancient practice of Winterage in rural Burren provides an opportunity for festivities. It reflects ancient Irish pagan festivals, originating in nature and in harmony with its people, farming, culinary, and local traditions (burren winterage.com; Dunford, 2016). Kim (2017), examining the Wine festival in Bordeaux, finds that festivals are about storytelling originating in a historical background benefiting nature. The Winterage festival is also rooted in storytelling, history, and traditions (burrenwinterage.com) and is a destination where place-based learning occurs [Burrenbeo.com(a), n.d.]. Place-based learning involves people learning in, about, and through place, engaging students in their community, including their physical environment, local culture, history, and people [Burrenbeo.com(a), n.d.; Minero, 2016]. The Winterage festival facilitates this learning [Burrenbeo.com(a), n.d.], and such learning can be found within edutainment activities.

Edutainment Education and Learning

Coined by Walt Disney in 1954, edutainment is defined as educational entertainment; education through entertainment (Sala, 2021). Gallucci, Bellelli, Saccà, and Addeo (2014) describe edutainment as a portmanteau from education and entertainment; it defines widespread activities on a continuum that goes from full entertainment forms to e-learning and multimedia systems. There is an expansion in cultural events (exhibitions, festivals, concerts) that constitute a form of event edutainment which can contribute to the intellectual growth of both the individual and society (Gallucci et al. 2014). Rossetti and Quinn (2019) found that travel and learning have always been interwoven and yet much scope remains for furthering understanding of the relationship between the two domains. The Winterage Festival discussed in this chapter echoes this narrative.

The Winterage Festival – A Case Study

The community-based Winterage Festival takes place annually, since its inception in 2014, on the October bank holiday weekend and celebrates ancient and iconic farming traditions [Burrenbeo.com(a), n.d.]. The BurrenBeo Trust was responsible for creating the festival with traditional age-old practices at its core. The timing of this festival is rooted in the old Irish folklore of the ancient harvest festival of "Samhain", marking the end of the harvest and the beginning of the winter. In ancient times, the festival of Samhain, the most important of the four Celtic festivals (Imbolc, Beltane, Lughnasadh and Samhain), marked the end of the Celtic year and the beginning of the new one (newgrange.com, n.d.).

The Winterage Festival mirrors the ancient ritual of winter grazing for cattle in The Burren. This rite is the stimulus for the festival that lasts for four days and, on the final day, cattle are moved from the lowland fields to higher ground of limestone rock which, over the summer, grew rich grasses and herbs; there they pass the winter grazing, benefitting from the warmth stored in the rock and drinking the calcium-enriched springs. This form of farming is based on sustainable methods of biodiversity and heritage and is vital in preserving the upper lands from reverting to vast areas of alder and hazel trees, bramble, and gorse. The concept of Winterage is an ancient system of transhumance (seasonal movement of people with their livestock between summer and winter pastures). This practice is an unusual form of "reverse" transhumance that finds cattle winter grazing outside in the elements on dedicated upper pastures; this is the antithesis of what typically happens in other transhumance systems and a phenomenon of The Burren (nationalinventoryich.chg.gov.ie, n.d.).

Methods

This study is exploratory where the concepts of learning, visiting, experiencing, and participating in a local festival in the west of Ireland are examined. The core examination and question centre on whether festival participants are aware that they are gaining education at this festival; do they attend the festival for entertainment and/or does education happen concurrently?

Addressing Rossetti and Quinn's (2019) view that there is much potential to extend research into the learning dimensions of festival experiences, a single case study approach is used specifically focusing on The Winterage. The case study method consists of designing a case study, collecting the study's data, analysing the data, and presenting and reporting the results (Yin, 2011). A case study is an "empirical inquiry about a contemporary phenomenon (e.g. a "case"), set within its real-world context – especially when the boundaries between phenomenon and context are not clearly evident" (Yin, 2009, p. 18). Yin (2009, 2011) relays how Bromley (1986) found case studies start from a desire to derive a close understanding of a single or small number of "cases" set in their real-world context, for example, The Winterage Festival, set in the natural landscape of The Burren. A case allows for closeness producing invaluable and deep understanding. This insightful appreciation of the case will result in new learning about real-world behaviour and its meaning (Yin, 2011), in this context, the Winterage Festival and edutainment.

This qualitative study fits the case study approach, where the research undertaken to explore this unique festival consisted of a multi-method approach consisting of desk research; autoethnography research; three in-person visitor journeys undertaken in 2018, 2019, and 2021; and interviews with key stakeholders, members of the Trust, custodians of the landscape and the farms, local enterprises, accommodation providers, and holidaymakers. Ten participants (see Table 9.1) were drawn from this pool to understand the concepts of the festival and interviewed in 2022. Prior to the interviews, an interview guide was developed based on relevant desk research, and from the researcher's visitor journey experiences. This interview guide was pilot tested with a trusted experienced tourism researcher. After cleansing the guide, a final improved guide was utilised. Once participants were identified based on their relevance, email communications were sent to request participation. Several key stakeholders were identified, but this was reduced to the most appropriate. Informed consent was requested. The study was explained in detail and all ethical issues outlined, including the right to be anonymous, voluntary participation, the right to withdraw and GDPR storage, and handling of data. All participants volunteered their time, and all agreed to have their data included in the study; some were happy to be named and others preferred to remain anonymous and are referred to as Participant Y, and Festival Attendees 1–3 – coded as FA1, FA2, FA3.

Table 9.1 Participants

Participant Name	Participant Code
Dr. Brendan Dunford, Manager, Burren Programme	BD
Áine Bird, Burrenbeo Trust Coordinator and Researcher	AB
Simon Haden, Managing Director, Gregans Castle, The Burren	SH
Sadie Chowen, Company Director, The Burren Perfumery	SC-BP
Trust Member	TM
Shane Casey, Cattle Drive facilitator and Farm Host	SC-CDH
Participant Y	Participant Y
Festival Attendees x 3	FA1, FA2, FA3
Short codes are applied in the findings.	

The interviews were recorded and transcribed by the researcher – where the first pass of the data was undertaken during transcription. The transcripts were read and reread several times to identify relevant rich content. Findings were coded and grouped into pertinent sections using colour coding as identifiers. Key findings emerged that produced rich informative data.

Findings and Discussions

The Winterage Festival

The Winterage Festival in The Burren culminates on the Sunday of the bank holiday with the Community Cattle drive. Overall, the festival consists of farm walks, talks and presentations, panel discussions, music, food tasting, educational events, and poetry reading, and finishes with the grand finale of herding the cattle to the upper lands, where locals, tourists, visitors, clergy, and dignitaries come together to hike up the trails after the cattle. The festival gives communities an opportunity to "engage with their own heritage" (AB). The Winterage provides a way to communicate to a wider audience about "high nature value" farming and celebrate all dimensions whether it is "food, heritage, sustainability, or landscape" (BD) and, recently, The Winterage has been recognised as part of Ireland's Intangible Cultural Heritage (ilri.org, n.d.). Findings here reveal that the strengths of the festival are that it exposes communities to each other, the farming community, the business community, and non-farming residents all benefit. As farming can be a solitary occupation, it is an opportunity to meet, discuss issues of the day, and see what the other "fellow is doing". A lot of the farmers work on marginal land, so the possibility of financial support derived from environmentally friendly practices is becoming more relevant to them.

The community is at the core of this region, and its festival and hosting the drive is an opportunity to contribute and engage with the community. Findings reveal that people are not as connected with or as understanding of farming as previous generations, and "this is a critical time when biodiversity is at the forefront of public policy" (SC-CDH). The farm host is central to the event; "anything we can do to raise the profile of the work Burren farmers do for conservation will support the continuation of The Burren Programme" (SC-CDH). The festival raises the profile of that unique farming tradition of Winterage and continues to attract national and international attention. It acts as a catalyst providing "an opportunity for people to visit, experience the landscape, the festival and learn the traditions from the community and its people" (TM).

Festival Origins

The festival originated from The Burrenbeo Trust, an independent landscape charity based in The Burren operating under the concept of place-based learning, namely, "learning about the place in the place for the place". The Burren Community Charter brought together a number of stakeholders: The Burren

IFA, The Burren and Cliffs of Moher, The GEO Park, (previously Burren connect), Burrenbeo Heritage Trust, and The Burren Life programme, (a conservation farming program). Spanning two counties, the heritage officer from Galway County Council and the heritage officer from Clare County Council tried to establish "what the people of the region wanted for the area" (AB). From a series of consultations, a charter was developed, resulting in positive actions, specifically the idea of a celebration for the unique farming traditions of The Burren. This gave birth to The Winterage Festival in 2012, starting as a small event and growing exponentially. The weekend starts with The Burren Winterage School, where official learning takes place along with sharing ideas around the farming in the Burren, called "high nature value farming", farming with nature. Originally ideas were only shared with farmers from Ireland, but now, people travel from all over the world to take part and to learn. This leads into the weekend, which is open to the public where the farm walks and farmer-led events occur, the sharing of specific "heritage and traditions" (AB) creating edutainment. The festival has evolved to include food events because farming and food are inextricably linked. On Sunday, the cattle drive takes place where hundreds of people turn up to a different farm and community every year. Community spirit is a key element and people gather to have a "cup of tea and scone" (AB), and then follow the farmers up to the uplands and the final journey of The Winterage. The festival highlights the wider value of the rural community and what visitors can learn from them. Importantly, "it celebrates" these people. The celebration "reflects on the local people who live in The Burren. It has a mirroring effect" (SC-BP).

The publication of the recent history of The Burren in 2002 from Brendan Dunford's Doctoral Thesis *Farming and The Burren*, highlights The Burren in terms of its flora as a result of farming practices. Contributing to this case, Brendan outlines that his job is working with farmers in *The Burren Life Programme*, a research project that was European funded and became The Burren Programme to work with farmers for "environmental landscape outcomes" (BD). Brendan found that farmers and environmentalists were not on opposite sides of the environmental argument. Farmers now claim that space, and this is very much Brendan's (and Burrenbeo Trust's) vision of how environmentally friendly landscape management must proceed. The Winterage festival is an important showcase for that concept; "indeed, one could say that reverse transhumance is the essence of landscape management in The Burren" (Participant Y).

Festival Community

Business contributors outline that they are fully committed to the region and the festival community, as one participant stated, "I was brought up here, it is of benefit to me and my business to see and understand the continuation of the traditional farming methods of The Burren and how they in themselves preserve the landscape" (SH). It is inspiring to hear contributors outline that they are part of a community; it's part of their identity, since of "pride of place" and complimentary to their business and sustainability. "We are founding members of The Burren Ecotourism Network, a group of tourist focused businesses and micro-business that adheres to an independently audited code of practice in sustainability" (SH). There is a sense of occasion about the cattle drive and supporting the farmer host is important. Participants are proud of their traditions and believe that the cattle drive sets them apart – it "reminds us that what we do in this place is different, special, and links us to the farmers that have gone before us" (SC-CDH). The festival is a contrivance, in that the practice of transhumance goes back thousands of years. The festival, gathering and re-enactment, and priest blessing of the cattle (see Figure 9.2) are new. "This has become a celebration of a singular piece of community history" (BD). The practice of moving cattle is commonplace all over the world and in the Bale Mountains of Ethiopia there is vertical transhumance similar to The Winterage called Godantu in which local populations move their livestock to the mountains in the winter when crops are being grown in the lowlands (ilri.org, n.d.).

This study finds that for the continuation of traditions and learning, future generations will be vital. Families make up a large part of the cattle drive with impressionable young children. "We mustn't forget children" (PY); they will secure the future of the region.

Benefits of the Winterage Festival

Participants strongly believe that The Winterage Festival adds value to the region while also raising awareness and providing learning for visitors about the environment, landscape, community, biodiversity, lifestyle, and heritage of

Figure 9.2 The cattle drive.
Source: (Burrenbeo).

The Burren. Businesses in the region outline that they notice an increase in visitor numbers around The Winterage Festival and in their business with one noting that "enhancing the guest experience and educating them about the wonders of The Burren, generates repeat business" (SH).

The benefits for the farming community is the validation that people are interested in what they are doing and are willing to learn from it. It adds an extra layer to the importance of caring for their traditions and place, whether they are directly farming or supporters within the local community in terms of preserving the biodiversity and archaeology of the region. The festival has been very successful in terms of the attention it has garnered all over the world. It connects communities around the globe that practise transhumance. The festival attracts many international visitors, particularly to the school which is part of the festival and where much of the learning happens in terms of sustainable farming – people talking about practical farming systems, methodologies, and practices. It provides the opportunity for researchers and policy makers to come together. "It's rare to have policy makers, scientists, researchers, students, practitioners, and farming advisers all in one room" (BD). For the festival attendees contributing, they find the immersion in the landscape and learning unparalleled and of course they are benefiting from edutainment. "I never miss a year of the cattle drive, the connection to nature and the landscape are second to none" (FA2). "The benefit for me is the immersion in the natural things in life and I always learn something new" (FA1).

The overall benefits are incalculable, and the connectedness is palpable. There is a feel-good factor to the event that must be experienced in person to be understood. The festival underlines other aspects outside of business, like human connectivity. The added value is that local businesses benefit mainly "due to the lovely atmosphere in the area and the life that the festival brings" (Participant Y). "The area is buzzing, and the benefits can be seen locally with bustling villages and happy people all immersed in the area. The benefits are immeasurable" (FA3).

Edutainment at The Winterage

Recognising The Burren as a learning landscape, the festival begins with The Winterage School, established in 2013. Mirroring Cabanas (2020), one participant notes, "It is a great learning event for farmers, but visitors that coincide with the festival learn about the ancient traditions by default" (AB). All contributors believe that the festival provides an opportunity for place-based learning. Place and emotions collide in a powerful way in this unique landscape with the learning landscape of The Burren providing the nucleus for learning, the festival the catalyst. The Winterage "is real and not 'put on', the public participate and learn" (AB).

Evidence-based learning happens at The Burren Winterage School. Transparent learning is provided for those interested in farming for nature of how farming can make a positive contribution to addressing the global biodiversity crisis through farm visits and walks, workshops, live events, and presentations. It endeavours to be as inclusive as possible, so participants and contributors develop an informed constructive approach to farming and the region through edutainment. Contributors are varied but all with a shared interest and common goals of sustainable farming (burrenwinterage.com). In 2021, for example, experts such as ILRI senior scientist Fiona Flintan, and Ann Waters-Bayer, the Agrecol Association for AgriCulture & Ecology (Germany), shared their experiences working in transhumance and rangeland systems in eastern Africa (ilri.org, n.d.).

When it comes to participants who attend for fun (Kim, 2017) and do not realise they are actively learning through edutainment, Brendan Dunford says, "If you ask me what I think people learn, I believe they learn that farming nature and heritage can exist together in harmony as distinct from the story that destructive farming is destroying the environment" (BD). For festival visitors, the unsuspected learning while holidaying in the region was the "turning on its head" (FA1) of normal farming practices. Cattle are normally housed in the winter and "not placed on top of a mountainous ridge" (FA1). "This was a total revelation for me, something that I never saw before" (FA2). The evidence is compelling that the visitors are experiencing edutainment and the next participant has the same experience. "I found it fascinating and totally enjoyed the immersed experience and the learning of moving with the heritage practices and seasons in this wild landscape" (FA3). "Who could have known in advance of visiting during this special time that I would learn of reverse transhumance" (FA2). According to Simon Haden, the evidence is there and his guests experience edutainment (Sala, 2021):

> Guests who participate in the festival are accidentally learning about the geology, natural history, botany, culture, social history and farming of The Burren and how this all weaves together to make this place unique. International tourists are usually here because of the Cliffs of Moher. They don't know anything about The Burren until they arrive.

Participants talk about the "natural learning experience" (TM). The evidence is that travel and learning go together (Gallucci et al. 2014), and this is supported here:

> Learning is an inherent part of travel, and I don't believe anyone visits The Burren without expecting to learn a little bit more about it. What might be unexpected is the community celebration. Where my own travels have overlapped with community celebrations, I always view it as a special bonus.
>
> (SC-CDH)

Providing an example of edutainment, this business owner believes that visitors to the perfumery as part of the festival see it through different eyes – the whole concept is holistic; "this is one of the major benefits". Rather than thinking "*oh, I have to learn this*, it opens people to learning as enjoyment. In the future, they will be open to learning about landscapes, biodiversity, sustainability" (SC-BP).

Findings in relation to the festival experience fit with Gallucci et al. (2014) that knowledge is created, shared, and enriched by the development of continuous learning and facilitated by cultural experiences. Aine Bird notes that festival participants

> get access to real experts of the landscape on the farm walks and access to private land. There's the National Park, the Green roads, but the vast majority is private. Being a guest of the landowner allows you to share their insights into people and place.

There are two levels of awareness in terms of learning by festival participants. "The, *oh right, you keep your cows on the mountain for winter*, and the more detailed understanding of what that involves for the farmer and the landscape" (SC-CDH).

The people of The Burren want visitors to enjoy and learn in a natural way in their natural setting. The festival is fun, and learning is part of it. Edutainment is for all. Brendan Dunford outlines here that he has no wish to "ram learning down people's throats", "we just let people enjoy it, and the farmer talks and answers questions. It's a hands-off approach in terms of how learning is communicated". The Winterage School has been the genesis of a lot of ideas and policies. Although The Burren programme is well established, it was The Winterage School that brought together farmers from home and abroad, drew scientists from Ireland and Europe, and encouraged the Department of Agriculture to invest in these new programmes.

Tourists engage in festivals because they find it enjoyable and educational (Rossetti & Quinn, 2019). Findings on The Winterage festival mirror this. In the context of The Winterage, the learning landscape model is important because, to appreciate the place and experience it fully, you need information.

> On the farm walk and cattle drive, information from the farmer of how the systems have evolved and how The Burren flora has developed, why the archaeology is protected, why The Burren Life Programme supports farmers to farm in a high nature value fashion adds so much to the experience. If you walk on your own with no background information, you will not have the same level of understanding and appreciation.
>
> (AB)

The festival is successful from the organisation's perspective in consolidating its position as an innovative educator. The festival is an opportunity to take part in an edutainment experience; "learning is an inherent part of any experience" (SC-CDH).

Conclusions

This is a learning landscape, a place where people come to learn.
– Brendan Dunford

Happiness of humans comes from the festival, the authentic play. The revitalisation of local festivals, which allow the increment of one's personal happiness is estimated to have the most crucial value (Kim, 2017). Even the least curious visitor who happens to arrive in The Burren during the festival will be tempted to ask questions; how, why, when? The local community answers, and, wrapped in the enjoyment of the tourist essentials of food, lodging, entertainment, one will discover a carefully crafted and immersive educational structure which addresses the topographical, historical, botanical, ecological, and extensive indigenous culture of this richly unique region. Led by farmers, the edutainment at The Winterage presents a "better insight into sustainable farming" (BD). The co-creation of knowledge with attendance in cultural experiences such as the festival facilitates lifelong learning (Gallucci et al., 2014).

This chapter explored through insightful contributions how this festival, and its interactive and co-creative approach to life, leads to the education of those visitors and tourists who are fortunate to visit. To answer the key questions outlined at the beginning of this chapter, experts willingly explored the concepts. Are festival attendees aware that they are being educated at this festival? The study revealed that education and learning is both active and passive, active through the school and passive as part of the edutainment experience. For those who actively "opt in" to the festival school, they know that they are there to learn and achieve, and for visitors who come to the festival for fun, they will be lucky enough to gain both entertainment and education in one unique suite; edutainment is here for the fortunate few. Education and entertainment are concurrent.

Enjoyment for participants, and a sense of place and purpose for the farmers, the local community, and the business owners, is clearly apparent. For many, learning is purposeful; they come to actively engage with the events; for others, it's passive, while attending a fun rural festival. Interest in The Winterage School is growing from other international farming communities who find that edutainment is bountiful: the ethos of the people, sense of belonging, community spirit, supporting each other, a natural affinity with nature, the learning landscape, and its strict protection are all entirely inspirational.

References

Bromley, D.B. (1986). *The Case Study Method in Psychology and Related Disciplines*, Wiley: Chichester.
Burrenbeo.com(a). (n.d.). *Burren Winterage Weekend.* https://burrenbeo.com/events/burren-winterage-weekend-3/
Burren.ie. (n.d.). *The Burren & Cliffs of Moher UNESCO Global Geopark.* https://burren.ie/

Burrenwinterage.com. (n.d.). *Burren Winterage School.* https://www.burrenwinterage. com/burren-winterage-school/

Cabanas, E. (2020). Experiencing designs and designing experiences: Emotions and theme parks from a symbolic interactionist perspective. *Journal of Destination Marketing & Management, 16*, 100330. https://doi.org/10.1016/j.jdmm.2018.12.004

Dunford, B. (2016). The Burren Life Programme: An overview. Research Series Paper No. 9, National Economic and Social Council Publications, Dublin, pp. 34–36. http:// burrenprogramme.com/wp-content/uploads/2019/10/NESC-Research_Series_ Paper_9_BDunford_Burren.pdf

Fáilte Ireland. (2019). *Fáilte Ireland announces funding of €3 Million for major festivals and events.* https://www.failteireland.ie/Utility/News-Library/Failte-Ireland-announces-funding-of-%E2%82%AC3-Million-for.aspx

Gallucci, C., Bellelli, P., Saccà, G., & Addeo, F. (2014). The assessment of cultural experience through the measurement of cross-cutting skills: The Giffoni experience case study. In L. Aiello (Ed.), *Handbook of research on management of cultural products: E-relationship marketing and accessibility perspectives* (pp. 83–112). IGI Global.

ilri.org. (n.d.). *Ancient farming practices in the Burren region of Ireland reveal the mutually beneficial relationship between farmers and the environment, International Livestock Research Institute.* https://www.ilri.org/news/ancient-farming-practices-burren-region-ireland-reveal-mutually-beneficial-relationship-between

Irishmyths.com. (n.d.). *Irish myths.* https://irishmyths.com/2022/02/01/what-is-imbolc/

Kim, K. S. (2017). Local festival and culture contents, *Journal of the Korea Convergence Society Korean Convergence Society, 8*(7), 183–189. https://doi.org/10.15207/JKCS. 2017.8.7.183

Minero, E. (2016). *Place-based learning: A multifaceted approach connecting learning to students' surroundings—the buildings, the landscape—covers content goals and builds community pride.* https://www.edutopia.org/practice/place-based-learning-connecting-kids-their-community

nationalinventoryich.chg.gov.ie. (n.d.). *Winterage in the Burren.* https://nationalinventory ich.chg.gov.ie/winterage-in-the-burren/

Newgrange.com. n.d. *Samhain (Samain) – The Celtic roots of Halloween.* https://www. newgrange.com/samhain.htm

Rossetti, G., & Quinn, B. (2021). Understanding the cultural potential of rural festivals: A conceptual framework of cultural capital development. *Journal of Rural Studies, 86*, 46–53. https://doi.org/10.1016/j.jrurstud.2021.05.009

Rossetti, G., & Quinn, B. (2019). Learning at literary festivals. In I. Jenkins & L. A. Lund (Eds.), *Literary tourism: Theories, practice and case studies* (pp. 93–105). CABI.

Sala, N. (2021). Virtual reality, augmented reality, and mixed reality in education: A brief overview. In D. Choi, A. Dailey-Hebert, & J. Estes (Eds.), *Current and prospective applications of virtual reality in higher education* (pp. 48–73). IGI Global.

Wright, A. S. (2009). Destination Ireland: An ancestral and emotional connection for the American Tourist. *Journal of Tourism and Cultural Change, 7*(1), 22–33. https:// doi.org/10.1080/14766820902807664

Wright, A. (2017). Heritage tourism and the curious tourist: The perspective of a small Irish town. *Tourism Today, 16*, 17–28, ISSN:1450-0906.

Wright, A. (2019). Collaborative learning: Businesses and HE co-create. *Learning Connections 2019: Spaces, People, Practice, 42*, 7–10. https://doi.org/10.33178/LC.2019.02

Yin, R. K. (2009). *Case study research: Design and methods.* Sage.

Yin, R. K. (2011). *Applications of case study research.* Sage.

10 Managing Inclusivity and Diversity with Edutainment for Festivals

Cemile Ece, Efnan Ezenel and Elif Şenel

Introduction

The term edutainment, which blurs education and entertainment, is primarily produced through visual materials, stories, and game-like forms (Aksakal, 2015). Edutainment encompasses activity processes involving many methodologies and visual media, and its primary goal is to connect people with information using technological, educational, and entertaining tools. By doing so, it grabs people's attention and maintains it at a specific level (McKenzie, 2000). The resources and messages provided in the edutainment process help individuals become more conscious of certain subjects while also learning via experience (Anikina & Yakimenko, 2015). Many organisations, including festivals, might have edutainment components as one of their goals. More recently, educational activities combined with entertainment have begun to take centre stage in festivals through various visual mediums (Chen, 2012). Documentaries and animations are examples of visual media tools utilised in education in these visual environments. In this context, edutainment plays an important role in film festivals attracting people's attention and raising awareness on a particular subject, such as equity and inclusion. As a result, in recent years, themes of equity and inclusion have become a part of the festival experience, and have played an essential role in raising social awareness at film festivals in particular. Inclusion and diversity are now at the forefront of festival design and are more widely addressed in events and organisations, which has had an impact on how organisations operate (Alvarado, 2022). Festivals that aim to be inclusive are organised for everyone and have emerged as the most powerful symbols of equality, diversity, and inclusion. As these themes have become more prevalent at film festivals, the demand for these festivals has grown even more as a result (Uzzo, 2020). The elements that a festival must have in order to integrate edutainment are discussed.

Information regarding the use of edutainment in film festivals is discussed in this chapter. Correspondingly, the topic of diversity and inclusion in film festivals is also covered through the use of a case study – the Accessible Film Festival. Interviews were conducted with attendees of Turkey's Accessible Film Festival, and information regarding edutainment events was obtained.

DOI: 10.4324/9781003305415-13

Edutainment (Educational Entertainment) in Film Festivals

Edutainment, in which education is supplemented by entertainment and technology, can include learning resources for both parents and children, to strengthen individuals' reading, writing, and thinking skills (Addis, 2005). Edutainment constitutes a process that configures knowledge and entertainment by adapting individuals' existing knowledge and intellectual skills to new information technology, raising expectations that learning may be fun (Colace et al., 2006; Salomon & Almog, 1998). Edutainment, which utilises the fewest words and most entertaining methods to transmit the most information, may also be encountered in entertainment media like computer games, multimedia software, websites, movies, and music. Simulation, graphic, and visual methods in education, particularly in the concept of edutainment, enable learners to comprehend and internalise difficult subjects (Charsky, 2010). People in the learning process adopt more creative ideas, increase their motivation, and become more participatory, thanks to Edutainment (Hussain et al., 2003). Edutainment is also carried out through workshops, camps, and festivals led by educational institutions as well as some private institutions, universities, and government institutions. Film, theatre, art, and education festivals, for instance, which are planned with entertainment and educational events, are a way to communicate emotions to the audience, such as happiness, melancholy, or excitement, with the objective of educating and training people (Seong-Hye, 2008). There may be differences in the length of festivals, even though the contents vary in those where a comfortable and cheerful environment is created. However, regardless of the tools and equipment utilised, all individuals can participate in the message or training to be provided.

The common goal of educational programmes at festivals is to establish an educational atmosphere, progress to the learning stage, and thus gain access to a set of tools to increase the level of awareness (Snell, 2005). A real or imaginary event, circumstance, or any other subject is animated and conveyed to the other party during these activities. Furthermore, the activities held as part of a festival can include trainings, meetings, projects, product launches, career planning, fairs, drama, theatre, cinema, short films, stand-up comedies, and performances with a variety of rich and inventive activities for all the senses of the participants (Harvey, 2003; Jarvin, 2015). During film festivals, a variety of information can be conveyed to the target audience via television shows, short films, serials, and documentaries, as well as educational entertainment via the screen (Kim, 2017). The festivals, which feature entertaining visuals and activities that encourage them to be memorable, also include simulation methods. These techniques enable technical training and demonstrations to be carried out for a person, including explaining how the festival occurred and what should be done as if it were occurring right now (Chismar & Brandman, 2014). When watching the experiences, all participants in the learning process have the ability to simultaneously enter the perception process, and it is crucial that both abstract and concrete interactions exist (Kuri, 2000).

Inclusivity and Diversity in Festivals

Innovative learning tools have been created and developed with the goal of inspiring people to adopt modern educational methods and materials in their didactic activities. Didactic events are those that are designed with the intention of educating, intending to transfer entertainment and pleasure while also imparting knowledge and teaching. To enhance learning, several schools and associations organise various didactic activities, such as excursions to laboratories, museums, and ancient cities, and culture days and diversity events for children and families. In addition, storytelling is a less-didactic way of addressing cultural issues (Esteban Núñez & Gómez Ríos, 2019).

Film festivals have also become part of educational entertainment as didactic festivals (Yrjänä, 2014). The annual International Short Films and Art on Illness Festival in Spain hosts digital artworks that attendees can view for one week. This film festival is free and accessible to all attendees and is thus one unique example of these festivals (Reigada et al., 2019). However, there are some physical and societal barriers that prevent some attendees from having equal access to and opportunity to participate in film festivals (Truter, 2018). Previously, when designing film festivals, the needs of disadvantaged groups, including but not limited to, people with different disabilities such as physical, and visual impairment or isolated elderly people and children, were not considered at all. The use of assistive technology or the absence of a sign language interpreter during the festival, the lack of education of the festival team on inclusive customer experience, and the lack of consideration for accessibility or accessibility of the venue when organising film festivals are some of the key challenges that can be mentioned. However, as clearly stated in Article 27 of the Universal Declaration of Human Rights – everyone has the right to freely participate in the cultural life of society, to benefit from art and to benefit from scientific progress and benefits (United Nations, 1948). This legal right is also closely linked to Goal 10 of SDG, which seeks to reduce inequality within, and between countries, by strengthening and promoting social, economic, and political inclusion of all, including persons with disabilities (Together 2030, 2019). Equal opportunities through film festivals are achieved by raising awareness of existing inequalities and creating policies, programs, and practices related to film and the festival to minimise them (Schwartz et al., 2010).

Reinforcing this, some film festivals create and apply various festival policies. Take One Action Film Festivals (TOA) created an Equality, Diversity, and Inclusion Policy, raising awareness against existing inequalities and ensuring inclusiveness and diversity. TOA incorporates equality, diversity, and inclusion into all its activities and working practices. This covers its primary organisational function as well as the festivals, events, projects, and opportunities to create, use, and advocate best practices. The main objective of this inclusive film festival in Scotland is to establish a secure and inclusive environment for all, regardless of their disability, financial status, race, gender identity, or sexual orientation. TOA encourages and supports networks' growing commitment to and awareness of equality, diversity, and inclusion. TOA film festivals are committed to the following policies (TOA Policy Plan, 2021):

- Equity: Actively work towards removing barriers to access for staff, volunteers, trustees, guests, audiences, and all contributors to events.
- Diversity: Ensuring that diversity is recognised, respected, and valued, supporting the representation of partially different experiences both on and off-screen.
- Inclusivity: Providing a practice and approach that enables people (audience, filmmakers, event contributors, staff and volunteers, the wider community) not only to be involved in activities but also to do so in circumstances they can determine for themselves.

Founded in 1988 and one of America's leading regional film festivals, the Virginia Film Festival (VAFF) also represents diversity in race, ethnicity, gender identity, age, physical ability, LGBTQIA+, and socio-economic background through its inclusive programs and festivals. It aims to ensure diversity, equality, and inclusion. Some of the VAFF's (2021) goals for diversity, equality, and inclusion included:

- Creating an annual Festival Program with balanced representation of gender and race, depicting the diversity of the human experience.
- To ensure meaningful participation and ownership in their programs by partnering with organisations and individuals representing different communities to better serve them.

In addition to the examples above, the Accessible Festivals Network "Accessible Film Festival, BE IN!" was established in 2019 at the initiative of Klappe Auf to enable attendees, individuals, disadvantaged groups, or individuals with any disabilities to watch films and participate in cultural life. Accessible Film Festivals (Erişilebilir Film Festivali) (Ankara, Turkey) are the film festivals that take place in the Accessible Film Festival BE In! (Disability Arts International, 2019). Accessible Film Festival, BE IN! aims to promote equal access to cinema-related activities and the representation of disadvantaged individuals in cinema and TV, with special opportunities for the audience with visual and hearing impairments and by bringing the issues of diversity, equality, and inclusion to the agenda at an international level (Esra, 2018). In this context, platforms such as Accessible Film Festival, BE IN!, VAFF, and TOA, while hosting accessible film festivals, aim to ensure diversity, equality, and inclusion in all parts of the festivals to all attendees.

Managing an edutainment film festival in the context of inclusion and diversity involves different dimensions. Equal opportunities must be provided for all in order for festivals to be inclusive and diverse (Oxoby, 2009). A successful film festival requires the management of these dimensions. Planning the entire process from start to finish, including all related risks, is required when organising a festival or event (Cudny, 2016). The most significant component of these materials is a well-informed team. Teams are the most important component for an organisation's success and management of the overall system (Watkins & Marsick, 1992).

The physical characteristics of the area, where the festivals are held, are also of great importance for the attendees. Having a suitable area, the sound system, the design of the space, the restrooms, lounge areas, and the layout of the event hall all affect the accessibility of the event (Ezenel, 2019). Providing accessible transportation is a crucial aspect of managing film festivals. In particular, in the locations where the festival will be held, the ease of transportation, the accessibility of public transportation, and the ability of people to plan their travels in a safe manner increase the demand for an event (Ilban & Kömür, 2019). The use of technology is also prevalent both before and during festivals in many different contexts. The use of accessible applications, audio descriptions, and smart ticket applications, particularly in festival promotion, facilitates the process (Kakon, 2022).

The Accessible Film Festival – A Case Study

The Accessible Film Festival has been chosen as a case study for this research because it places a high priority on diversity and inclusion. This research will determine the management methods used and uncover the key criteria needed. The festival screens its films with Turkish audio descriptions and Turkish caption subtitling in an effort to create a more accessible festival.

Outstanding international and Turkish films are shown to the audience within the festival's framework in a variety of topic categories. The festival, which began in Ankara, has been meeting with moviegoers in Istanbul and Eskişehir since 2017. Parallel events such as Q&A sessions with filmmakers, workshops, virtual reality experiences, and exhibitions are also presented by Accessible Film Festival in order to broaden the cinema offer and culture.

Methodology

Grounded theory has been a common method in qualitative research since the 1960s (Pranskūnienė, 2017). This method enables researchers to carry out analysis in studies (Ece, 2019). Grounded theory is applied to produce new theories, offer varying perspectives, and design new frameworks (Sarı et al., 2020). However, it has only been applied as a method in this study, not as a means of developing theory. This research utilises grounded theory because it enables the researcher to investigate further, to go beyond the limits of constructing or applying theory, and to conceptualise the experiences of the research participants (Ezenel, 2019).

The topic of research has been how diversity and inclusion are managed for edutainment at festivals. The Accessible Film Festival, which prioritises diversity and inclusion, has been chosen as a case study research area in order to determine this management process and to reveal the key criteria used.

Interviews were conducted with festival organisers and attendees to analyse the entire process, from festival organisation to participant inclusion. A total of 27 people (six organisers and 21 attendees) who took part in these events between 2018 and 2021 were interviewed. The interviews were conducted using open-ended questions. Apart from the open-ended interviews, participants in the event were first asked to describe the day they attended the festival. These stories were taken in written form, and the interviews were recorded. The records were then deciphered and analysed. It was agreed upon by the participants that their voluntary involvement would be subject to ethical regulations for data gathering. The confidentiality of personal data was given priority.

Grounded theory is a method for inductively structuring theory extraction within the context of a research topic (Çelik & Ekşi, 2015). Glaser (1992) notes grounded theory is "a general methodology of analysis linked with data collection that uses a systematically applied set of methods to generate an inductive theory about a substantive area". In-depth analysis then enables the researcher to develop new concepts and conduct required comparisons for similarities and differences between these concepts (Buluklu, 2019). As a result, grounded theory contributes to the illumination of new and understudied research topics – such as equity and inclusion in film festivals. The Straussian grounded theory approach developed by Strauss and Corbin (1998) was adopted in this study. This is because, as part of the grounded theory analysis process, the researcher is allowed to interpret and conceptualise the data through analysis and categorise these concepts in relation to existing literature to reveal new findings (Strauss & Corbin, 1990).

Findings

The analysis of online and face-to-face conversations within the scope of COVID-19 measures led to the formation of certain categories. Inclusion and diversity are regarded in these categories as additional requirements for an edutainment festival. It is also seen how an edutainment festival such as The Accessible Film Festival can be managed along with these categories. The categories are described in further detail below.

Organisation

Findings from The Accessible Film Festival indicate that the primary activity in the development and implementation of festivals is organisation. During the interviews, the festival organisers discussed this topic, stating,

> Our event, The Accessible Film Festival, appealed to a certain segment of the population, and we included disadvantaged groups in our team so that we could understand all the situations they might need, so that we could not only organise for disadvantaged groups but also an organisation with them.

This was confirmed by another participant who stated, "Friends in the organisation team helped us a lot.... They could understand us, they could understand our needs". The team needed members from disadvantaged individuals, individuals with special needs, and minority groups in order to be able to empathise, raise awareness about what might actually be needed, and provide a good service, especially at festivals where inclusiveness and diversity are held in high regard. Marketing is another element of an organisation, and it is critical to choose marketing channels based on the target audience. When the promotional channels for the Accessible Film Festival for disadvantaged groups are examined, it is clear that accessible technology methods are implemented. For example, Participant 3 from the organisation team stated: "For the Accessible Film Festival, we chose social media tools that everyone could use and reach, and the most important point in this was that we preferred applications with accessibility features".

A smart planning strategy involves creating a checklist of the necessary items. Knowing the priority area proficiently and recognising the gaps and requirements are essential for effective planning (Frost & Laing, 2015). Therefore, it is important to conduct preliminary audience research while planning the festival. On the aforementioned issue, Organiser 4 stated:

> Before holding the festival, we wanted to develop a good strategy by meeting with our team, especially the issue of how the organisation would be organised was extremely important for us. For this, we did our research as a team. We created our needs list and organised our organisation according to our needs list.

It is thus necessary to plan for the needs of the target group in order to ensure inclusion and diversity. Making appropriate designs and planning for the individuals who will participate at every level of the organising is also another thing that must be done for the planning to be inclusive and appropriate for diversity in edutainment festivals. To achieve this, the festival planners should follow an inclusive design process when creating the technological and educational tools that will be utilised in the events and activities, which is further confirmed stated by Organiser 2 who commented,

> We generally use different techniques such as gamification or storytelling in the educational activities we do at our festival. Our main point is to have fun while learning. While doing these, we make our games inclusive for the participants of the activities we design.

Transportation/Location

Effective transportation is one of the most significant components in the organisation of a film festival to encourage a high participation rate. Regarding transportation, Participant 2 noted, "It is extremely important for us that the event is in the centre. I was able to come to the organisation hall easily with my wheelchair and public transportation". This was echoed by Participant 6 who stated,

> It is difficult for us to go to events in long distances.... I have to go by private car because I am pregnant and that causes someone to bring me in the process.... When there are events that are close and I can go on my own, I can come alone.

In review of these comments, it is clear that having a comfortable event and an accessible location is significant for high participation.

Physical Environment

The area where the festival is held is accessible, has adequate manoeuvring space, and has all the required infrastructure. Therefore, the level of participation is high. Regarding this issue, the organisers explained,

> It is extremely important that everyone can move freely in the festival area, even if we choose a hall for an organisation, it is a priority for us that it is comfortable, accessible in every aspect and its hosting capacity is a priority for us.

This was recognised by the participants, in which Participant 13 commented, "I can act alone, and it is extremely important for me to reach an upstairs hall without any help". Drawing on these statements, it is therefore important that the selection of the festival venue has taken into consideration the ability to provide attendees with a free space. In addition, wheelchairs and walkers should always be available to guarantee accessibility at the festival. Planning for activities within the festival venue is another primary factor and should be planned according to the participation of the individuals, their disability, and the location of the stage. On this point, the organisers observed, "We planned a seating arrangement close to the sound system for visually impaired individuals". Participant 10 also commented, "I requested to sit closer to the exit door due to my allergies. Since I was affected by a lot of dust and smoke, the officers at the festival helped me with this".

Social Inclusion

The Accessible Film Festival is free, to encourage participation of individuals with financial barriers. In relation to this, Participant 14 remarked, "It is extremely important for us that the events are free. As a retired person, unfortunately, it is difficult for us financially to participate in all events". Furthermore, it was also determined that individuals felt excluded in activities just organised for a particular group. According to Participant 15, there are feelings of exclusivity, which they commented,

> When I attend an event, people look at me as if I am different, which causes me to leave the event immediately. However, I do not feel marginalised at all events where I am greeted with a smiling face by the people who embrace everyone and welcome me.

Technology

The materials used during the festival must be accessible and applications combined with technology should be included. It is necessary to create audio descriptions for the visually impaired, use instant translation software, and implement inclusive practices in seminars and training held throughout the event (Dalaslan & Şulha, 2019; Rahman et al., 2022). Regarding this, Participant 12 expressed, "We expected the questions to be repeated several times in the event, and we could not hear some of the questions". Participant 7 also remarked, "We are glad that there are audio descriptions.... Otherwise, we are waiting for our friend next to us to explain or help us". This correlates with Anikina and Yakimenko (2015), who stated that edutainment is a significant step forward from traditional education methods, with modern technology merging entertainment and education.

Content/Methodology

A festival should be planned with a fundamental purpose in mind and should be designed accordingly (Hertzman, 2006). Likewise, the content of the festival determines the methodology of the materials, seminar, or training to be selected. Learning through experience sharing is one of the most common methods in edutainment activities. As a result, the method should be chosen in accordance with the message to be delivered at the festival, the awareness to be raised, or the subject to be taught. Assistance with content selection should come from well-informed professionals. Participant 18 noted, "We see a problem that we have experienced in all the selected films or an incident that we witnessed, observed.... Or we see a story that encourages us, and that is very important to us".

Communication

The most significant tool in the fulfilment of an event and organisation is communication. The communication language that is utilised must also be inclusive and unifying if we are to talk about inclusivity in the fulfilment of an event. Organiser 1, for example, commented, "We paid great attention to the language we use, and we left aside the words that are still used and marginalising in the society....We even trained all our staff for this". During the workshops and other programmes held at the festival, language assistance needed to be available. Notably, participant profiles and their needs should thus be considered during planning. Language assistance should include sign language, translation, simultaneous interpretation, and audio descriptions. On this topic, the participants explained, "The situation we expect most from the events is language support, some of us do not have English and sometimes we have difficulty in reading the subtitles, we would like this problem to be considered". Language difficulties are covered under communication and are experienced by people with disabilities that have disabilities on hearing, speaking, reading, writing, and/or comprehension, and who communicate in different ways than other people. Thus, language is a crucial barrier to participation in events and the inclusion of people with disabilities in the events.

Conclusion

All the categories needed for inclusion and diversity in festivals were identified after analysing the data and findings. As a result of the interviews, these categories were examined and discovered using the grounded theory methodology. Examining the categories reveals that they are fundamentally human-oriented, with all the complexities disclosed to produce an inclusive model (see Figure 10.1).

A strategy might be discussed that emphasises inclusiveness and diversity from the planning stage onwards, particularly in festivals that have an entertainment component. Building a team is the first step in the inclusion process as the team is crucial to the development and operation of any firm (Jepson & Clarke, 2014). Inclusion is at the forefront, and the team is shaped accordingly,

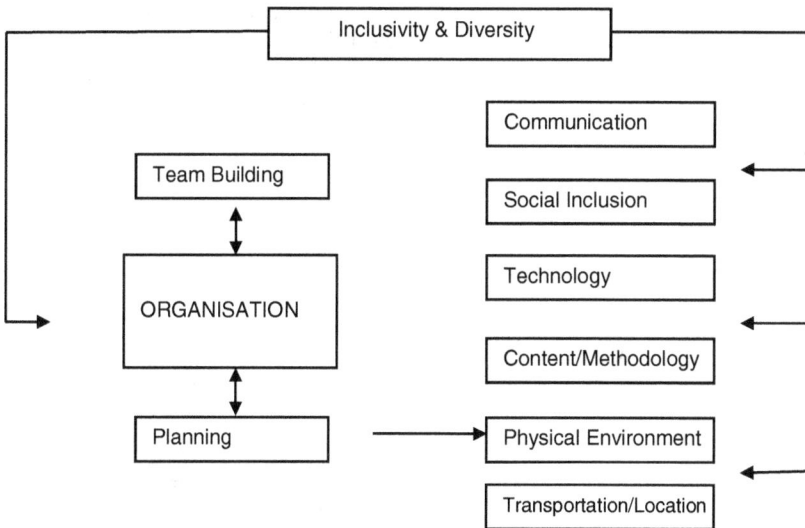

Figure 10.1 Inclusive model of edutainment festival.

as was noted in the findings. This is crucial in determining the needs, realisation, promotion, and completion of the organisation. The need for greater knowledge and empathy with the team members is essential.

The adoption of a structural inclusion and diversity program is invaluable. This process includes everything from the physical environment to the materials used, from the technological infrastructure to transportation and placement. To enable people to engage without difficulty, it is vital to take these difficulties into consideration. The resources that were utilised to create the content, training, and seminars must be available. Additionally, the event's terminology should be inclusive and unified. On the other hand, employing technology to enhance education is one of the most popular strategies in edutainment activities. By incorporating strategies like gamification, technology, and storytelling, the development of the educational environment should increase participant motivation (Hussain et al., 2003). In addition, edutainment activities must be human-centred or user-centred. For this reason, it is necessary to design and develop the profiles of the participants, their needs, and the structure of the organisation in the planning stages (Walldén & Soronen, 2004).

In summary, it is necessary to adopt an inclusive, facilitating, and unifying practice in matters related to organisation, transportation/location, physical environment, social inclusion, communication, content, and technology, in order to ensure inclusiveness and diversity in an edutainment festival.

References

Addis, M. (2005). New technologies and cultural consumption – Edutainment is born! *European Journal of Marketing, 39*(7/8), 729–736. https://doi.org/10.1108/0309056051 0601734

Aksakal, N. (2015). Theoretical view to the approach of the edutainment. *Procedia – Social and Behavioural Sciences, 186*, 1232–1239. https://doi.org/10.1016/j.sbspro. 2015.04.081

Alvarado, K. G. (2022). Accessibility of music festivals: A British perspective. *International Journal of Event and Festival Management, 13*(2), 203–218. https://doi.org/ 10.1108/IJEFM-12-2020-0082

Anikina, O. V., & Yakimenko, E. V. (2015). Edutainment as a modern technology of education. *Procedia – Social and Behavioural Sciences, 166*, 475–479. https://doi. org/10.1016/J.SBSPRO.2014.12.558

Buluklu, Y. (2019). Embedded theory as a qualitative research method in critical studies. *Journal of Critical Communication Studies, 1*(1), 1–14.

Çelik, H., & Ekşi, H. (2015). *Qualitative patterns: Embedded theory*. Express Press.

Charsky, D. (2010). From edutainment to serious games: A change in the use of game characteristics. *Games and Culture, 5*, 177–198. https://doi.org/10.1177/15554120093 54727

Chen, L. C. (2012). A study on the interaction styles of an augmented reality game for active learning with a folk festival book. In M. Soares & F. Rebelo (Eds.), *Advances in usability evaluation part II* (pp. 113–122). CRC Press.

Chismar, W. G., & Brandman, A. (2014). Summer film festivals for kids: Building fond memories and brand identity. *Summer Academe, 8*, 2–9.

Colace, F., De Santo, M., & Pietrosanto, A. (2006). Work in progress: Bayesian networks for Edutainment. *36th Frontiers in Education Conference*. https://doi.org/10.1109/FIE.2006.322573

Corbin, J., & Strauss, A. (1990). Grounded theory research: Procedures, canons, and evaluative criteria. *Qualitative Sociology*, *13*, 3–21. https://doi.org/10.1007/BF00988593

Cudny, W. (2016). *Festivalisation of urban spaces*. Springer.

Dalaslan, D., & Şulha, P., (2019). Accessible films festival as an accessible event space for the visually impaired. *Journal of Translation Studies and Applications*, *27*, 64–88.

Disability Arts International. (2019). *Be In! An International Accessible Film Festival Network*. https://www.disabilityartsinternational.org/resources/be-in-an-international-accessible-film-festival-network/

Ece, C. (2019). *A study on ecovillages in the scope of sustainable tourism: The case of Turkey* [Master's thesis, ESOGÜ, Sosyal Bilimler Enstitüsü].

Esra, Z. (2018). *Festival aims to make films accessible for the disabled*. https://www.dailysabah.com/cinema/2018/10/12/festival-aims-to-make-films-accessible-for-the-disabled

Esteban Núñez, M. T., & Gómez Ríos, A. V. (2019). Storytelling as a didactic proposal to approach culture topics in a fifth graders' classroom. *Cuadernos de LingüísticaHispánica*, *33*, 169–187.

Ezenel, E. (2019). *A study on inclusive tourism: The case of Eskişehir* [Master's Thesis, ESOGÜ, Sosyal Bilimler Enstitüsü].

Frost, W., & Laing, J. (2015). Avoiding burn out: The succession planning, governance and resourcing of rural tourism festivals. *Journal of Sustainable Tourism*, *23*(8–9), 1298–1317. https://doi.org/10.1080/09669582.2015.1037841

Glaser, B.G. (1992). *Basics of grounded theory analysis*. Sociology Press.

Harvey, D. (2003). *The state of postmodernity*. Metis Yayınları.

Hertzman, E. (2006). *Visitors' evaluations of the historic content at Storyeum: An edutainment heritage tourist attraction* [Doctoral dissertation, University of British Columbia].

Hussain, H., Embi, Z. C., & Hashim, S. (2003). A conceptualized framework for edutainment. Informing Science: InSite-Where Parallels Intersect, 1077–1083.

İlban, M. O., & Kömür, T. (2019). The role of festival tourism in destination branding: The example of international olive harvest days in Ayvalık. *Travel and Hotel Management Magazine*, *16*(2), 274–295.

Jarvin, L. (2015). Edutainment, games, and the future of education in a digital world. *New Directions for Child and Adolescent Development*, 147, 33–40. https://doi.org/10.1002/cad.20082

Jepson, A., & Clarke, A. (Eds.). (2014). *Exploring community festivals and events*. Routledge.

Kakon, K. (2022). Technological adaptation in tourism events, fairs, and festivals: Way to a revolutionary transformation in Bangladesh. In A. Hassan (Ed.), *Technology application in tourism fairs, festivals and events in Asia* (pp. 167–180). Springer Singapore.

Kim, K. S. (2017). Local festival and culture contents. *Journal of the Korea Convergence Society*, *8*(7), 183–189.

Kuri, N. P. (2000). Kolb's learning cycle: An alternative strategy for engineering education, *Educational Resources Information Center*, pp. 3–7. https://eric.ed.gov/?id=ED441666

McKenzie, J. (2000). Beyond edutainment and technotainment from now on. *10*(1). http://www.fno.org/sept00/eliterate.html

Oxoby, R. (2009). Understanding social inclusion, social cohesion, and social capital. *International Journal of Social Economics*, *36*(12), 1133–1152. https://doi.org/10.1108/03068290910996963

Pranskūnienė, R. (2017). Grounded theory methodology in the context of social innovations for rural development research. *International Scientific Conference Rural Development*, 1273–1279. http://conf.rd.asu.lt/index.php/rd/article/view/416

Rahman, M. K., Rana, M., Akter, S., & Hassan, A. (2022). Technology innovation as a destination branding tool for festivals and events tourism. In A. Hassan (Ed.), *Technology application in tourism fairs, festivals and events in Asia* (pp. 135–147). Springer.

Reigada, C., Martín-Utrilla, S., Pérez-Ros, P., Centeno, C., Sandgren, A., & Gómez-Baceiredo, B. (2019). Understanding illnesses through a film festival: An observational study. *Heliyon*, *5*(8). https://doi.org/10.1016/j.heliyon.2019.e02196

Salomon, G., & Almog, T. (1998). Educational psychology and technology: A matter of reciprocal relations. *Teachers College Record*, *100*(1), 222–241. https://doi.org/10.1177/016146819810000201

Sarı, Y., Pranskūnienė, R., Ece, C., & Ezenel, E. (2020). Experiences at ESOGÜ tourism camp: A grounded theory approach. In I. O. Coşkun, N. Othman, M. Aslam, & A. Lew (Eds.), *Travel and tourism: Sustainability, economics, and management issues* (pp. 331–342). Springer.

Schwartz, D., Blue, E., McDonald, M., Giuliani, G., Weber, G., Seirup, H., ... & Perkins, A. (2010). Dispelling stereotypes: Promoting disability equality through film. *Disability & Society*, *25*(7), 841–848. https://eric.ed.gov/?id=EJ906110

Seong-Hye, A. (2008). Strategic planning for initiating local cultural festivals. *The Journal of the Korea Contents Association*, *8*(12), 168–175. https://doi.org/10.5392/JKCA.2008.8.12.168

Snell, K. (2005). Music education through popular music festivals. *Action, Criticism & Theory for Music Education*, *4*(2), 2–35.

Strauss, A., & Corbin, J. M. (1990). *Basics of qualitative research: Grounded theory procedures and techniques*. Sage Publications.

Strauss, A., & Corbin, J. (1998). *Basics of qualitative research: Techniques and procedures for developing grounded theory* (2nd ed.). Sage.

Take One Action Film Festivals (TAO). (2021). *TOA Policy Plan: Equality, diversity and inclusion policy*. https://www.takeoneaction.org.uk/wp-content/uploads/2021/07/TOA_EDIPolicyPlan-2021-23_revised.pdf

Together 2030. (2019). *Realizing the SDGs for all: Ensuring inclusiveness and equality for every person, everywhere*. https://sustainabledevelopment.un.org/content/documents/23216Together_2030__Position_Paper__HLPF_2019.pdf

Truter, L. (2018). Breaking ground-disability rights a reality in the film industry. *TFM Magazine*, *2*(14), 52–55. https://journals.co.za/doi/abs/10.10520/EJC-eca91c3b2

United Nations. (1948). *Universal Declaration of Human Rights: Article 27-1*. https://www.un.org/en/about-us/universal-declaration-of-human-rights

Uzzo, G. (2020). Accessible film festivals: A pilot study. *Bridge: Trends and Traditions in Translation and Interpreting Studies*, *1*(2), 68–85.

Virginia Film Festivals (VAFF). (2021). *Diversity, Equity, and Inclusion Committee*. https://virginiafilmfestival.org/wp-content/uploads/2015/09/Virginia-Film-Festival-Diversity-Equity-and-Inclusion-Committee.pdf

Walldén, S., & Soronen, A. (2004). *Edutainment – From television and computers to digital television.* (Tutkimusraportteja – Working Papers, No. 2). http://www.uta.fi/hyper/julkaisut/b/fitv03b.pdf

Watkins, K. E., & Marsick, V. J. (1992). Building the learning organisation: A new role for human resource developers. *Studies in Continuing Education, 14*(2), 115–129. https://doi.org/10.1080/0158037920140203

Yrjänä, A. (2014). Film festivals as effective tools in media education. Tampere University of Applied Sciences. https://www.theseus.fi/bitstream/handle/10024/77429/Yrjana_Aino.pdf;jsessionid=8A7156391F842F76644A29B10D04DEFB?sequence=2

11 Edutainment Actions in a Mexican Film Festival

Cinema Planeta, the First Film Festival in Morelos, Mexico

Driselda-Patricia Sánchez-Aguirre

Introduction

The theme of sustainability has become a topic of interest in various spheres of society, with education being a fundamental pillar for its promotion and, above all, to act in supporting it. Therefore, education and awareness-raising on environmental issues are becoming more and more urgent to assimilate and take on as a society, as we find our impact on the planet, day by day, is amounting to climate change, pandemics, humanitarian issues, and food and political crises – all which are increasingly compromising the existence of life as we know it, as well as that of other species with which we cohabit.

There are various efforts focused on integrating more and more people in the achievement of actions for sustainable development, one of them being the UN Sustainable Development Goals (SDGs). In 2015, the UN Member States adopted the SDGs as an open call for action to advance prosperity, people, partnership, peace, and planet. Known as the five Ps of sustainable development, they encompass the 17 goals of the agenda (Visser, 2015). Although many of the SDGs appear to be ambitious, one of the first steps to achieving them is to identify and raise awareness of them; thus, sustainability education is crucial to their achievement (Venske, 2021). And although the goals are global, the problems, as well as their possible solutions, are unique to each territory. For example, in Mexico some of the actions that have been taken in relation to sustainability education generally arise through government with public policies that address the issue from different aspects, such as preserving the knowledge of native peoples (Government of Mexico, 2018) or contributing to the separation of waste (Municipality of Querétaro, 2017). Civil society, private companies, and NGOs have also raised their voices through various movements (Rodríguez Cardozo, 2018). A result of these movements has been the development of festivals, including film festivals, which are sometimes organised by the private sphere, sometimes by NGOs or by civil society or a combination of all the three.

Although there is a wide range of academic literature on film festivals and sustainability (Blumenfield, 2021), and a few mentions on the relevance of edutainment and festivals (Viljoen et al., 2018), there is scarce academic

DOI: 10.4324/9781003305415-14

research about film festivals in Mexico specifically, and its relationship with sustainability education. One of the film festivals that have been taking place for more than ten years in the centre of the country with the theme of the environment and sustainability is Cinema Planeta. This festival is important because in addition to serving as a recreational space, it has an important social, cultural, economic, and environmental impact that people value, which, according to Getz (2008), contributes to make a festival transcendental. By exploring the impact of Cinema Planeta and its use of edutainment activities to maintain audience engagement, this chapter enhances the existing literature and extends current understanding of Mexico's contributions to sustainability initiatives.

Literature review

Traditional educational experiences are generally associated with one or a few people who speak or act while many others listen or spectate (Tongtep & Boonlamp, 2021). By involving the use of technology with entertainment and education, the term "edutainment" emerged (Rapeepisarn et al., 2006), and became known as "the amalgamation of education and entertainment" (Viljoen et al., 2018). Although there are different types of edutainment, it generally moves from a passive learner to active learners who usually participate in the generation, acquisition, and transfer of knowledge. According to Rapeepisarn et al. (2006), edutainment can be classified according to locationality, which can be interactive/participatory or non-interactive/viewing; according to the target group, who may have similar interests or the same age; and by the type of media used, which can be television, computer, tablets, etc.

In films and television shows, the tendency is towards non-interactive location-based edutainment, where the attendee plays the role of spectator to the content projected on the screen (Rapeepisarn et al., 2006). Even with passive roles, it has been proven that media content influences behaviour (Goldstein & Usdin, 2008). Some films and entertainment programmes have adapted their content to emphasise certain social behaviours in a way that inspires viewers to imitate them, driving behavioural change (Arabi & Benkhlafa, 2022; DellaVigna & La Ferrara, 2015). Although in some cases, there is no specific behavioural change associated with film watching (DellaVigna & La Ferrara, 2015), the use of static images like photos, drawings, or illustrations or moving images as in the movies or TV shows, has proven to be useful to reach the conceptualisation of social phenomena (Whitley, 2013).

In digital media, social media is a fundamental part of successful edutainment, and it is a virtual space where it is possible to measure engagement (Isacsson and Gretzel, 2011). However, most edutainment approaches in digital environments have shown limited potential for engagement (Frossard et al., 2015), perhaps because of a lack of precise description of the variables that make up the concept and hence its practical operability.

Film festivals use the benefits of image projection to captivate their audiences on a variety of themes, and like most festivals are familiar with adapting to unforeseen situations (Coscia & Voghera, 2022). But in early 2020, the COVID-19 pandemic left many organisers damaged by the situation and the multiple restrictions they faced (Nhamo et al., 2020). Among the sanitary restrictions applied at the international level, it was widely recommended to stay at home, postpone travel, and maintain social distancing (UNWTO, 2020) because there was a high possibility of contagion in closed spaces. People working in the organisation of festivals were therefore forced to take their work to the virtual world, and social networks gained wide relevance as a vital means of communication. Most festival organisers in 2020 had to hastily move to the digital environment to try to save their revenues or to maintain their contact with the public (Volpi et al., 2020), and many of the environments designed for offline learning migrated towards edutainment.

Isacsson and Gretzel (2011) identify certain characteristics of edutainment content design for social networks such as attractive visual material, game-like format, interaction, emotions, and setting a learning objective. While they present an application of these elements in a case study on tourism students, there is neither clear description of each element nor their possible link to engagement.

Discussing attractive visual material, Mestre (2012a) suggests with the current lifestyle, there is a continuous exposure to multiple types of information simultaneously making it difficult to process it optimally, using graphics or static visual elements becomes effective for learning as they demand less short-term memory. If the visual elements include a variety of colours and anthropomorphic figures, they become more attractive and easier to remember (Shangguan et al., 2020). Some studies have shown that the use of images in learning environments increased students' understanding of key issues and directed them towards reflective learning (Whitley, 2013). Game-like formats, however, have been described as content designed with a gaming approach, which involves challenges that, when solved, provides players/learners with a sense of accomplishment (Qian & Clark, 2016). Game-based learning thus positively influences learners' attitudes and cognitive achievement (Connolly et al., 2012). Rahimi et al. (2022) identified that meaningful learning through games occurs when there is immediate feedback, clear goals, varied rewards, and low-risk failure.

On the other hand, interaction has been described as content designed under an edutainment approach, involving a high degree of interactivity, which when applied by festival organisers is widely linked to engagement. Engagement can thus be understood as perceived interactivity and shared emotions (Yang et al., 2010). Most definitions of engagement coincide in appealing to three elements: cognitive, which alludes to mental or referential constructions towards a brand, object, or organisation, in this case towards a festival; emotional, which suggests emotions play a very relevant role in the development of experiences (Rodríguez-Campo et al., 2022; So et al., 2014); and behavioural, which refers to the actions that consumers take regarding the festival they are "following", ranging from sharing or commenting on experiences online (Lewis et al., 2021). Within the scope of the emotional element, it is important to note there is also a scale of emotions applicable to festivals, which identifies the major categories of love, joy, surprise, and negative emotions (Lee & Kyle, 2013). Finally, setting has been described as a learning objective immersed in these actions. In virtual environments, it is feasible to include a variety of learning objectives to make the teaching–learning cycle flexible, interactive, and focused on the segment targeted by the learning strategy (Mestre, 2012b).

Cinema Planeta

In 2009, through a civil association, Cinema Planeta, the first film festival in Morelos, Mexico, was held. Since then, the festival's main objective has been to broadcast short films and movies associated with the UN SDGs encouraging critical thinking among the audience. Its audience varies between rural and urban inhabitants with a wide age range from children to senior citizens.

The festival is divided into three parts: 1) the academic part which lasts 15 days and has an average of 25,000 attendees each year; 2) the film festival with green carpets (that replaces the usual stereotypical red carpet at film festivals), premieres and guest film directors which lasts six days and has gathered up to 60,000 attendees; and 3) a tour around Mexico which lasts six months. The festival takes place in Morelos, Mexico, and takes place in 12 venues within the city. When the festival goes on tour, there are an additional 30 venues nationwide.

The key networks with which the festival collaborates on a frequent basis are varied and include, for example, the National Council of Science and Technology in Mexico, Earth Day Network, Green Film Network, and Worldwide Environmental Education. Thanks to these networks, it has been possible to maintain a constant sponsorship that has strengthened the festival in various aspects, one of which is the educational line of the festival.

Cinema Planeta has developed a series of activities that go beyond the screen, allowing the audience, especially the younger ones, to participate in various activities related to the film they have just watched. In this way, and in a face-to-face manner, with few resources associated with technology, Cinema Planeta commonly links entertainment with learning. For example, when the audience is younger, at the end of the film the festival organisers do an activity related to what the children have just seen. For example, if the film's emphasis was underwater life off the Mexican coast, the children are given paper, scissors, and colouring tools to make a representation of what appeared in the film, such as a whale.

When the audience is older, such as teenagers or adults, and if the directors or producers of the film are present, after the film projection, there is a question-and-answer session. When there are no people involved in the making of the film after the film projection, the festival organisers take the initiative to ask questions randomly that invite the audience to reflect on what they have recently seen on the screen.

Methods

To identify edutainment practices implemented by Cinema Planeta during the pandemic, two actions were carried out: a) in-depth hour-long interviews with the co-director and founder of the festival in April 2020, followed by a second series of interviews in 2021 to update the festival's data. The interviewee agreed to be mentioned in the research outcomes as Cinema Planeta's Director (CP Director), and b) measurement of engagement on Facebook through quantitative analysis of reactions and comments to images published during 2020–2021 in the biography of Cinema Planeta's fan page. Facebook was selected as the social network of interest since it is the most used social network in Mexico, second only to WhatsApp (Statista, 2022).

The total number of images published in the study period was 487 (43% of 2020 and 57% of 2021). From these, only those that had at least one reaction, a comment, or were shared at least once were selected. The final sample included 171 images. A content analysis was conducted using the sample images and interview transcripts. Content analysis is a method frequently used in social sciences (Camprubí & Coromina, 2016) and is defined as the interpretation of a particular corpus of visual, audible, or written material to identify patterns, biases, and meanings (Berg, 2007).

The presence or absence of elements alluding specifically to the characterisation of edutainment content mentioned by Isacsson and Gretzel (2011) and defined in the literature review were identified, and the emotional and behavioural elements of engagement were identified as part of the interaction.

Results and discussion

The results are presented in the order of elements identified with the content analysis based on Isacsson and Gretzel's (2011) classification of edutainment elements.

Visual materials

In general, the images published during both periods coincide in having mostly blue, green, and white colours, which are associated with harmony, hope, and goodness (Heller, 2008), which is very congruent with the philosophy of environmental care. The images of 2021 show people gathered with the use of face masks, which evidences the realisation of a hybrid festival. This is unlike 2020, where many of the images are drawings with anthropomorphic figures or allude to the covers of the films to be presented.

Game-like formats

To the surprise of many festival organisers, the founder and main organiser of the film festival considered the pandemic as a trigger for innovation within her festival. "Cinema Planeta is not cancelled and not postponed, it is adapted and transformed" (CP Director, 2020). Unlike many other festivals in which the digitalisation of activities caught them unaware, Cinema Planeta had already contemplated, but not established, to carry out part of its festival virtually, so the idea of adapting it to digital media was easier than in other festival cases. Game-based learning activities however had to be adapted to the digital environment. From the organisation's perspective, the results obtained from these actions were beneficial for the festival and were only possible because of the networks with which the festival had previously collaborated.

In the 2020 edition, a trivia game was implemented, which was promoted through social networks and carried out through the Idea Planeta streaming platform. The game for film viewers consisted of multiple-choice questions related to the film viewers just watched on the platform. Feedback was immediate, which Rahimini et al. (2022) identified as important in successful game-learning formats. When the challenge was accomplished, the sense of achievement that Qian and Clark (2016) mentioned consisted in that those who got all the answers right won extended time to access the entire offer of short films and films of the festival.

In review of the digital festival 2020, the director of Cinema Planeta identified, "We did a survey and 75% of the people who were on the platform did not know cinema planeta before it was digitalised. We quintupled [our presence]. We had 240,000 visits to the platform, we were in 193 cities" (CP Director, 2020). Because of this success, the 2021 edition was held in a hybrid format with both online games and face-to-face activities.

Interaction

According to the interviewee, during 2020 and the first stage of the pandemic, maintaining interaction with the audience was possible, thanks to the sponsorship of the Science Council that provided funding for the platform, as well as working in partnership with other organisations that contributed to disseminating and sponsoring the content of the festival.

> We had a lot of feedback on the films and people really liked the games, that was the result of a team effort we made and the support of fellow organisations and sponsors who, thanks to our good reputation, remained committed to the cause.
>
> (CP Director, 2020)

She also mentioned that in 2021, the face-to-face activities had an excellent response, especially from the children who seemed to enjoy learning through games. She commented, "It was lovely to hear a child saying to his parents, *We should take shorter showers to save the planet*, just after he played with clay to build a story about a world without water" (CP Director, 2021).

In Facebook communication, emotional and behavioural factors were identified. Emotional factors were identified as the Facebook possible reactions to an image (Merril & Oremus, 2021) and were classified as positive ("like", "love"), negative ("sad, "angry"), and neutral ("haha", "wow"). Behavioural factors were identified as the number of times the selected images were shared and the number of comments about them.

The highest peaks related to behavioural factors, specifically with sharing publications, are identified in 2020 (see Figure 11.1), highlighting April, when the total digitalisation of the festival is announced. Interaction through comments is very low in both periods. However, the images with the most comments in both periods of study coincide in being invitations to participate either as volunteers (2020) or as spectators (2021) to the Festival's opening ceremony of the hybrid festival. In 2020, the image was accompanied by text mentioning the virtual games that would be available during that festival. Both of the images can be categorised as call-to-action stimuli (i.e. explicit invitations or suggestions to perform specific behaviours). It has been shown that someone is more likely to respond to this call if the content has some meaning for the viewer (Oltra et al., 2022), in this case, the attractiveness of online games was an important trigger.

Regarding emotions, reactions peaked in April 2021, the month in which it was announced that the festival would take place in a hybrid format, perhaps because activities outside of cinema theatres were expected after a long time of confinement. Most of the reactions identified were positive or neutral (95%) in both study periods. Negative reactions, such as sadness, were mainly related to an image showing a female face along with a text referring to a renowned Mexican film journalist's obituary. It is easy to understand why it was the image with the most negative reactions in 2020, but these negative emotions were not related to edutainment-related visual or text posts.

Learning objectives

In Cinema Planeta 2020 edition, entertainment was encouraged by being able to access films and short films from the comfort of home, and, as in previous editions, the learning objectives were based on the UN SDGs. Both formats, online (2020) and hybrid (2021), had a section of tips presented on screen at the end of the film, consisting of advice on activities to be carried out at home to contribute to the achievement of the SDG that was immersed in the projected film. For example, the CP Director commented,

> If you have a film that talks about water, at the end of the film we project 10 actions that you can do to help with the issue of water or rubbish or deforestation, so don't worry, you'd better get busy.
>
> (CP Director, 2020)

Although film screenings are per se an applied use of technology, in both 2020 and 2021, Cinema Planeta decided to continue with other activities that include the internet to complement technology. They used a video streaming platform and social media communications, which were especially important when

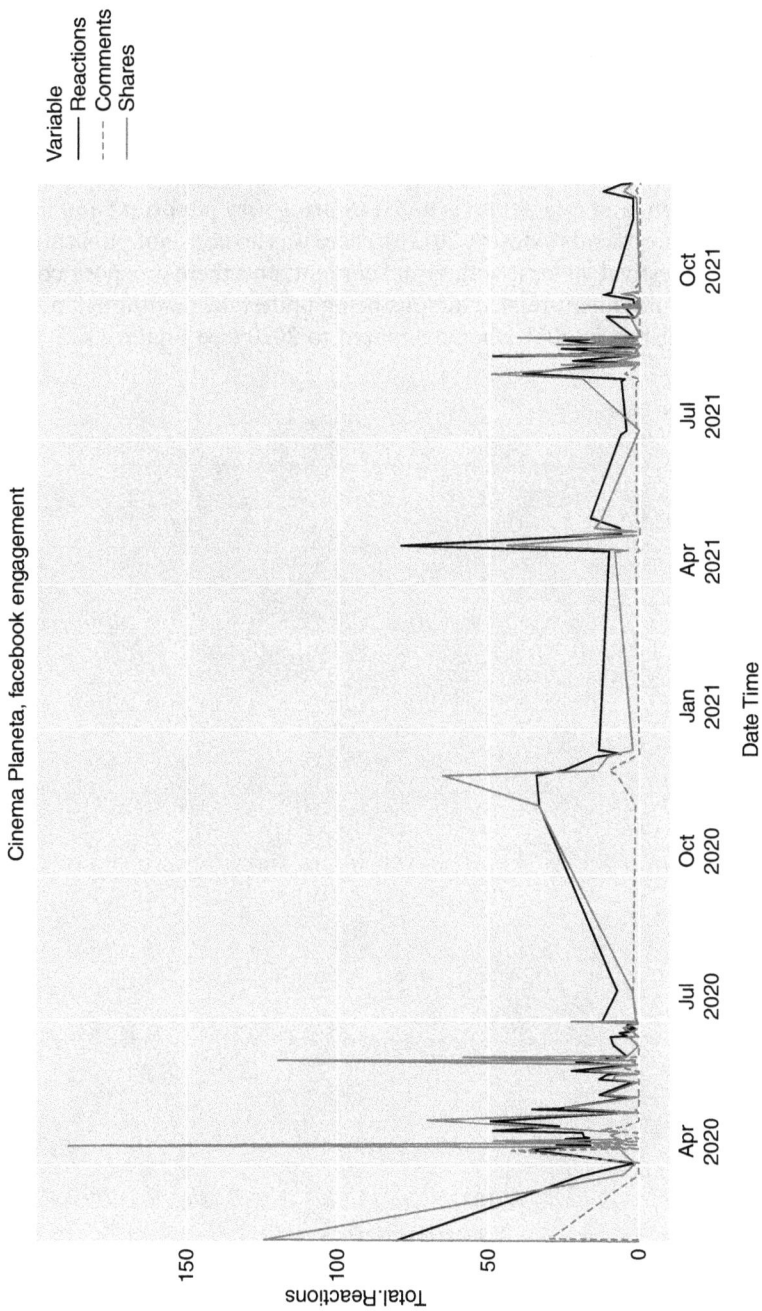

Figure 11.1 Facebook's engagement by month/year.

Developed by the author.

trying to attract a segment of digital natives (Isacsson & Gretzel, 2011) who are immersed in a streaming culture (Arditi, 2021; Neira et al., 2021). In the 2021 edition, with the hybrid edition, some of the offline educational activities were community film clubs or workshops about crafts produced by children. Therefore, technology became an important factor for the festival, which the CP Director noted: "Through technology, we celebrate examples, we motivate people to change" (CP Director, 2021).

Although Facebook was also a key factor to raise awareness of the festival in its digital edition, and interaction is noted as being very important and flexible in virtual environments (Mestre, 2012b), there was no constant publication schedule for the festival, at least with visual content, and there was poor communication of edutainment-related actions being undertaken. Although more images were published in 2021 when compared to 2020 (see Figure 11.2), the

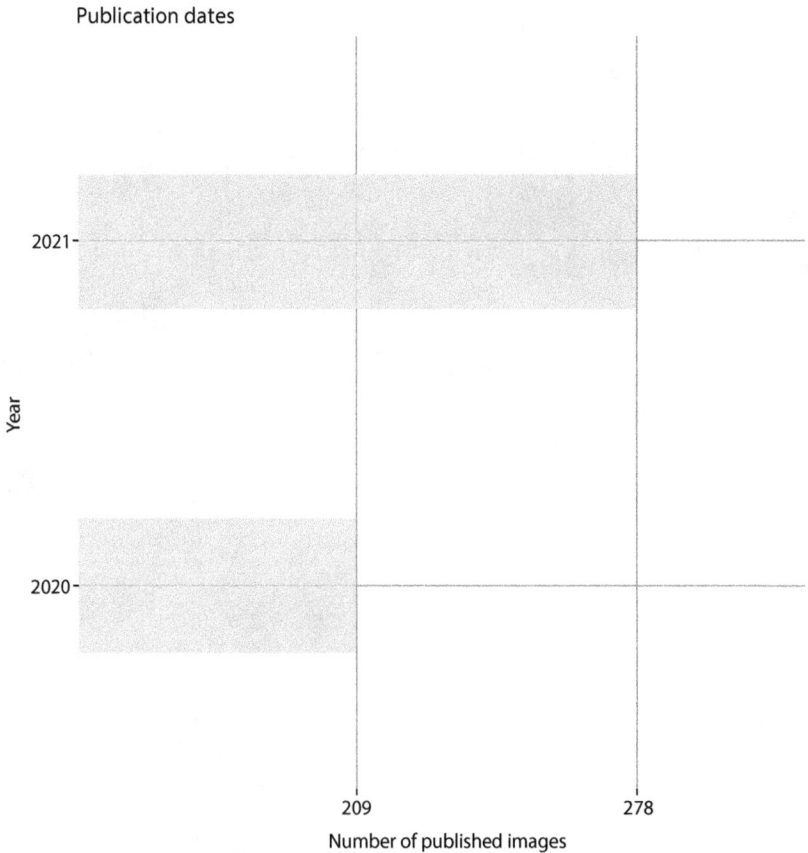

Figure 11.2 Publication dates.
Developed by the author.

publication dates were more constant in the first year of the study, (i.e. better distributed, and their engagement indicators were much higher, which suggests that the audience is looking for updated spaces and constant publication to interact with them).

Conclusion

Through this case study, it was possible to describe actions carried out by the Mexican film festival Cinema Planeta, also highlighting the importance of collaborative networks to maintain engagement with the audience. Cinema Planeta has been considered one of the few film festivals that take place in Morelos, Mexico and have a national scope. With play activities and advising texts related to the environment and the objectives set by the UN, the festival tries to influence the behaviour of spectators and volunteers to contribute to the sustainable development of the environment. Therefore, several techniques related to play and learning had been applied year after year with different audiences. With the changes that occurred due to the COVID-19 pandemic, in 2020, the festival was completely digitised, and previously established play-learn processes and activities were adapted to digital media to reinforce the sustainability issues communicated in the films.

It is understandable that, as this is the first time that Cinema Planeta has used digital or hybrid events, the adaptation of its game-learning activities has certain limitations, among them – the lack of communication of these activities and their results on Facebook, or even the use of this network to carry out the activities themselves, given that it is the social network reported as the one most used by the organisation.

The study was limited only to the analysis of images on Facebook and interviews with organisers of a single festival, it would be worth exploring differences and similarities in the use of edutainment activities beyond the mere screening of films between the case study presented in this chapter and other Mexican film festivals not focused on the environment, such as the Morelia Film Festival, the Guanajuato International Film Festival, the Durango International Film Festival, among others. However, the case study presented here is a clear example of how through simple actions that include the basic elements of edutainment (Isacsson & Gretzel, 2011) a festival that has been focusing on film-related play-learning activities for years can then adapt its activities to virtual environments. The actions identified arise as a consequence of adversity, a situation that has been documented to enable festival adaptation and innovation (Van Winkle & Kullman, 2022).

It would be interesting to identify more precisely and achieve concrete descriptions of specific edutainment actions focused on behavioural change linked to the UN SDGs through film, the research path is open for further studies to strengthen the theme through diverse methodological approaches such as experimentation or ethnography.

References

Arabi, H., & Benkhlafa, H. (2022). Le premier festival du film en Afrique et dans le monde arabe qui traite des questions de handicap: Un appel à l'action. *Motricité Cérébrale, 43*(3), 115–117. https://doi.org/10.1016/j.motcer.2022.07.002

Arditi, D. (2021). *Streaming culture: Subscription platforms and the unending consumption of culture.* Emerald Group Publishing.

Berg, B. (2007). An introduction to content analysis. In B. Berg (Ed.), *Qualitative research methods for the social sciences* (pp. 238–267). Allyn and Bacon.

Blumenfield, T. (2021). Fostering campus-wide dialogue and student-centered learning through film festivals and media projects: Engaging Chinese environmental issues beyond the Asian studies classroom. *ASIANetwork Exchange: A Journal for Asian Studies in the Liberal Arts, 27*(2), Article 2. https://doi.org/10.16995/ane.305

Camprubí, R., & Coromina, L. (2016). Content analysis in tourism research. *Tourism Management Perspectives, 18*, 134–140. https://doi.org/10.1016/j.tmp.2016.03.002

Connolly, T. M., Boyle, E. A., MacArthur, E., Hainey, T., & Boyle, J. M. (2012). A systematic literature review of empirical evidence on computer games and serious games. *Computers & Education, 59*(2), 661–686. https://doi.org/10.1016/j.compedu.2012.03.004

Coscia, C., & Voghera, A. (2022). Resilience in action: The bottom up! Architecture festival in Turin (Italy). *Journal of Safety Science and Resilience.* https://doi.org/10.1016/j.jnlssr.2022.10.005

DellaVigna, S., & La Ferrara, E. (2015). Economic and social impacts of the media. In S. P. Anderson, J. Waldfogel, & D. Strömberg (Eds.), *Handbook of media economics* (Vol. 1, pp. 723–768). North-Holland. https://doi.org/10.1016/B978-0-444-63685-0.00019-X

Frossard, F., Trifonova, A., & Barajas, M. (2015). Teachers designing learning games: Impact on creativity. In G. P. Green & J. C. Kaufman (Eds.), *Video games and creativity* (pp. 159–183). Academic Press. https://doi.org/10.1016/B978-0-12-801462-2.00008-4

Getz, D. (2008). Event tourism: Definition, evolution and research. *Tourism Management, 29*, 403–428. https://doi.org/10.1016/j.tourman.2007.07.017

Goldstein, S., & Usdin, S. (2008). Children, media and health. In H. K. Heggenhougen (Kris) (Ed.), *International encyclopedia of public health* (pp. 648–656). Academic Press. https://doi.org/10.1016/B978-012373960-5.00446-9

Government of Mexico. (2018). Beneficios de las políticas públicas para los Pueblos Indígenas. http://www.gob.mx/epn/articulos/beneficios-de-las-politicas-publicas-para-los-pueblos-indigenas

Heller, E. (2008). *Psicología del color: Cómo actúan los colores sobre los sentimientos y la razón.* Gustavo Gil.

Isacsson, A., & Gretzel, U. (2011). Facebook as an edutainment medium to engage students in sustainability and tourism. *Journal of Hospitality and Tourism Technology, 2*(1), 81–90. https://doi.org/10.1108/17579881111112430

Lee, J. (Jiyeon), & Kyle, G. T. (2013). The measurement of emotions elicited within festival contexts: A psychometric test of a Festival Consumption Emotions (FCE) scale. *Tourism Analysis, 18*(6), 635–649. https://doi.org/10.3727/108354213X13824558188541

Lewis, N., Palmer, A., & Asaad, Y. (2021). Linking engagement at cultural festivals to legacy impacts. *Journal of Sustainable Tourism, 29*(11–12), 1810–1831.

Merril, J., & Oremus, W. (2021, October 26). Five points for anger, one for a 'like': How Facebook's formula fostered rage and misinformation. *Washington Post.* https://www.washingtonpost.com/technology/2021/10/26/facebook-angry-emoji-algorithm/

Mestre, L. S. (2012a). 8 – Interactivity options for tutorials (pp. 171–203). https://doi.org/10.1016/B978-1-84334-688-3.50008-1

Mestre, L. S. (2012b). 4 – The need for learning object development. In L. S. Mestre (Ed.), *Designing effective library tutorials* (pp. 55–75). Chandos Publishing. https://doi.org/10.1016/B978-1-84334-688-3.50004-4

Municipio de Querétaro. (2017, April 11). Municipio de Querétaro inaugura la primera papelera monumental. *Municipio de Querétaro.* https://municipiodequeretaro.gob.mx/municipio-de-queretaro-inaugura-la-primera-papelera-monumental/

Neira, E., Clares-Gavilán, J., & Sánchez-Navarro, J. (2021). New audience dimensions in streaming platforms: The second life of Money Heist on Netflix as a case study. *Profesional de la información, 30*(1), Article 1. https://doi.org/10.3145/epi.2021.ene.13

Nhamo, G., Dube, K., & Chikodzi, D. (Eds.). (2020). Implications of COVID-19 on gaming, leisure and entertainment industry. In *Counting the cost of COVID-19 on the global tourism industry* (pp. 273–295). Springer International Publishing. https://doi.org/10.1007/978-3-030-56231-1_12

Oltra, I., Camarero, C., & San José Cabezudo, R. (2022). Inspire me, please! The effect of calls to action and visual executions on customer inspiration in Instagram communications. *International Journal of Advertising, 41*(7), 1209–1234. https://doi.org/10.1080/02650487.2021.2014702

Qian, M., & Clark, K. R. (2016). Game-based learning and 21st century skills: A review of recent research. *Computers in Human Behavior, 63*, 50–58. https://doi.org/10.1016/j.chb.2016.05.023

Rahimi, S., Shute, V. J., Fulwider, C., Bainbridge, K., Kuba, R., Yang, X., Smith, G., Baker, R. S., & D'Mello, S. K. (2022). Timing of learning supports in educational games can impact students' outcomes. *Computers & Education, 190*, 104600. https://doi.org/10.1016/j.compedu.2022.104600

Rapeepisarn, K., Wong, K., Fung, C., & Depickere, A. (2006). Similarities and differences between learn through play and edutainment. *Proceedings of the 3rd Australasian conference on interactive entertainment*, 4–6 December 2006, Perth, W.A., 28–32.

Rodríguez Cardozo, L. (2018). El desarrollo de las ONG de México y su coincidencia con los Objetivos para el Desarrollo Sostenible de Naciones Unidas. *CIRIEC-España, Revista de Economía Pública, Social y Cooperativa, 91*, 59. https://doi.org/10.7203/CIRIEC-E.91.8879

Rodríguez-Campo, L., Alén-González, E., Antonio Fraiz-Brea, J., & Louredo-Lorenzo, M. (2022). A holistic understanding of the emotional experience of festival attendees. *Leisure Sciences, 44*(4), 421–439. https://doi.org/10.1080/01490400.2019.1597790

Shangguan, C., Wang, Z., Gong, S., Guo, Y., & Xu, S. (2020). More attractive or more interactive? The effects of multi-leveled emotional design on middle school students' multimedia learning. *Frontiers in Psychology, 10*. https://www.frontiersin.org/articles/10.3389/fpsyg.2019.03065

So, K. K. F., King, C., & Sparks, B. (2014). Customer engagement with tourism brands: Scale development and validation. *Journal of Hospitality & Tourism Research, 38*(3), 304–329. https://doi.org/10.1177/1096348012451456

Statista. (2022). *Redes sociales más populares en México 2022.* Statista. https://es.statista.com/estadisticas/1035031/mexico-porcentaje-de-usuarios-por-red-social/

Tongtep, N., & Boonlamp, L. (2021). Educator personality toward edutainment for preparing youth to a digital society. *2021 2nd SEA-STEM International Conference (SEA-STEM)*, 25–29. https://doi.org/10.1109/SEA-STEM53614.2021.9668049

UNWTO. (2020). UNWTO Remains At a Standstill as 100% of Countries Impose Restrictions on Travel. World Tourism Organization. https://www.unwto.org/news/covid-19-world-tourism-remains-at-a-standstill-as-100-of-countries-impose-restrictions-on-travel

Van Winkle, C. M., & Kullman, S. (2022). Remaking the festival business model during a pandemic. *Event Management*, *26*(6), 1335–1350. https://doi.org/10.3727/1525995 22X16419948390989

Venske, E. (2021). Quality education: Industry contributions to embed sustainability in a meeting and event management curriculum. *International Journal of Event and Festival Management*, *12*(3), 297–313. https://doi.org/10.1108/IJEFM-12-2020-0079

Viljoen, A., Kruger, M., & Saayman, M. (2018). The art of tastings: Enhancing the arts festival experience. *International Journal of Event and Festival Management*, *9*(3), 246–265. https://doi.org/10.1108/IJEFM-03-2018-0023

Visser, W. (2015). *5 Ps of sustainable development: UN sustainable development goals*. http://www.waynevisser.com/report/sdgs-finalised-text

Volpi, J., de la Torre, G., Flores, J. I., González Ávalos, L. F., de la Garza Toledo, E., & Nivón Bolán, E. (2020). *Para salir de terapia intensiva. Estrategias para el sector cultural hacia el futuro*. Universidad Nacional Autónoma de México. http://www.librosoa.unam.mx/handle/123456789/3074

Whitley, C. T. (2013). A picture is worth a thousand words: Applying image-based learning to course design. *Teaching Sociology*, *41*(2), 188–198. https://doi.org/10.1177/0092055X12472170

Yang, S.-U., Kang, M., & Johnson, P. (2010). Effects of narratives, openness to dialogic communication, and credibility on engagement in crisis communication through organizational blogs. *Communication Research*, *37*(4), 473–497. https://doi.org/10.1177/0093650210362682

Part IV
Experiencing Edutainment

12 Edutainment through Storytelling, Nostalgia, and Living History in 1940s Festivals and Events

Zoe Leonard and Julie Whitfield

Introduction

This chapter provides a critical review of edutainment as provided through storytelling, nostalgia, and living history within 1940s festivals and events. This chapter captures the 1940s period by providing a review of how 1940s festivals and events are immersive, entertaining, and educational for the public and underpin a nuance of respect and remembrance. Inherently, as the 1940s generation declines, so too do the real stories and living memories of the time (Kirka, 2022); therefore, this chapter identifies the importance that future generations keep this historic legacy alive. The authors conclude that 1940s festival organisers have a cultural duty to present an accurate representation of the past, facilitating both education and cultural connections in an edutaining capacity.

Literature review

Cultural duty to present an accurate representation of the past

The 1940s was an era dominated by the war effort, the early 1940s were consumed by the largest conflict in history. This played a significant role in the development of life and living conditions within the UK during the latter part of the decade (Thompson, 2019). At the end of the war, crowds gathered across London, on Whitehall to hear Churchill speak, and in front of Buckingham Palace to witness King George VI address the nation with thanks for its efforts. Celebrations erupted nationwide, street parties saw bunting strung across the streets, with music playing as everyone celebrated the end of the war (Imperial War Museum, 2022). The jubilation was short lived as the focus then moved to life after the war, mass destruction of homes, the loss of over six million lives, and the realisation that life had changed forever (Hodge, 2012), an uncertain future lay ahead.

World War II (WW2) persists in engaging the minds of the British people (Hill, 1999). Almost all people will have a connection to the time – a relative

DOI: 10.4324/9781003305415-16

who enlisted as a boy, a mother or grandmother who worked in the ammunition factories, or a great-grandfather who protected the Homefront. It was a significant event that impacted the entire nation. The people of Britain were the heroes in their own story, the "people's war" (Calder, 1992). It is an event which shaped the thoughts and actions of those who lived many years afterwards. For the British, the memories of the war create a sense of a nation (Hill, 1999). The WW2 generation displayed a level of patriotism and fearlessness rarely seen. They sacrificed their own lives to enable the stability, security, and safety of future generations. In her diaries as part of a government initiative of mass observation, Nella Last indicated that even though she knew her son could come to his death, she would not stop him from playing his part in the war effort (Last, 1939 in Broad & Fleming, 2006). This collective nature, comradery, and spirit in the face of true fear are what individuals of today search for in their connection to the past – an element of the 1940s that modern communities would like to keep and rejuvenate. A certainty that came in everyone working together for the benefit of the nation. Morale was kept high through government propaganda, often silencing real stories of fear in favour of the British stiff upper lip (Noakes, 2014). This element of propaganda may lead one to question how much of the collective memory is selective, excluding the unsavoury elements of fear (Noakes & Pattinson, 2014) to champion a story that generates national pride and a more romantic memory of the time (Hodge, 2012). Thus, 1940s-inspired festivals and events endeavour to encapsulate these elements of the period and have a cultural duty to present this information as accurately as possible. Telling the story through the implementation of edutainment, to educate and inform, often generating a socio-cultural connection with the past. The entertaining inclusions of 1940s-inspired festivals and events aid in maintaining relevance to a modern audience, bolstering the immersive and informative experience in a fun and engaging way.

Edutainment – the synergy of education and entertainment through immersion

The concept of edutainment is based on the combined effect of providing education via an entertaining means (Santonen & Faber, 2015), often presented through storytelling connected to a specific theme. Edutainment can provide an immersive and/or passive experience (Wyatt, 2022) for audiences. In the case of 1940s-inspired festivals and events within the UK, attendees are exposed to an immersive and holistic encounter of the era (DeGroot, 2008), which provides an understanding of the past through remembrance and celebration (Getz & Page, 2016). The combination of historical re-enactment, themed entertainment, music, dancing, fashion, sounds, and smells incite an emotional engagement (Cho et al., 2014), often evoking a sense of nostalgia, a cultural connection, and an atmospheric link to the past (Carneiro et al., 2019). For some, these festivals and events are a direct memory of the time. For others, they enable an understanding of historic events. In either case, an element of imagination is required to allow an interpretation of what it would have been like in the 1940s. As González-González et al. (2022) assert, this can be supported through narration and re-enactment, bringing history alive.

Wagner (2015) stated that edutainment stimulates the historical imagination and enables the visitor to emotionally connect to the past. However, there has been some concern regarding edutainment within heritage settings, questioning if it is possible to effectively educate in an entertaining way, without overshadowing the importance of historical learning. Packer and Ballantyne (2004) contest this notion, instead finding that combining education and entertainment within a leisure setting provides a greater effective, engaging, and memorable experience. This is supported by Wyatt et al. (2021), who assert that settings providing edutainment experiences can entertain attendees whilst simultaneously educating on a specific historic time.

Storytelling

Stories of the past are relied upon within families and communities as a powerful tool for entertainment but also to educate on the past (The History Press, 2022). Storytelling facilitates a reminder of strength and survival within communities (Huber et al., 2013). Fog et al. (2005) assert that storytelling is a narrative technique for expressing everyday occurrences. In the case of the 1940s, we hear stories of the Homefront, the blitz, make do and mend, mass evacuation, and stories from veterans highlighting the traumatic experiences of those fighting for their country (Hodge, 2012) – "It has become a narrative of national identity, a story we tell ourselves about ourselves" (Hill, 1999, p. 324).

Storytelling within festivals and events facilitates an efficient exchange of information through social interaction (Moscardo, 2020). Festivals and events inspired by the 1940s can place themselves at the forefront of providing the narrative of the WWII era, keeping history alive for generations to come. The combination of people, objects, photos, a particular place, and stories within these festival and event experiences can create an authentic storytelling experience. Jones (2019) considers the strength of storytelling to be such that it engages the audience and invites them to become a part of the story itself. As seen in promotions for 1940s-inspired festivals and events, organisers often call for attendees to dress in authentic or 1940s-inspired outfits (Railway in Wartime Weekend, 2022). The attendees thus become a part of the story, transported back in time for an interactive and immersive storytelling experience. Zhou et al. (2012) discuss how socio-cultural stories through families and communities can also spark elements of nostalgia, strengthening the connectedness to the past.

Nostalgia

Nostalgia is defined as a positive yearning for the past (Chhabra, 2012). Nostalgia can often be generated through smells, tastes, sounds, or objects, sparking an idealistic recall and a desire to be removed from a less-than-satisfying present (Cho et al., 2014). Gao et al. (2020) identify two types of nostalgic experience: historical nostalgia – to remove oneself from a modern time and return to a more appealing past, and personal nostalgia – reflecting on personal experience and positive memories of a specific time within one's life. The generation with first-hand experience of the 1940s is slowly diminishing, taking their knowledge of a time gone by with them. One may question how it is possible for the next generation and beyond to have a nostalgic experience without first-hand knowledge of the era. However, it is not necessary to have first-hand knowledge of the time to enable a nostalgic experience. Indeed, one may draw upon knowledge generated through books, films, TV, museums, and education (Gao et al., 2020), through a sort of vicarious nostalgia (Goulding, 2002), along with stories passed down through the generations, and a community based on culturally collective memory. The level of authentic satisfaction experienced during an historical re-enactment event or visit to a place of heritage can directly impact upon the strength of nostalgia one encounters (Gao et al., 2020). Simultaneously, nostalgia can itself be the emotional element that motivates a visit in the first instance (Prayag & Del Chiappa, 2021).

Living history and re-enactment

Living history may be experienced through re-enactment of a specific time reinforcing education (Rodriguez, 2016) and entertainment through the provision of immersion. This type of experience offers a simulation of the past through a combination of immersive participation and spectatorship. In living history, authenticity can be deemed as a reflection of how real the historical representation is or seems. It is a marker for the research, knowledge, and detail that the re-enactor has contributed (Knowles, 2016) as they are fully immersed into a representation of the past. Living history provides an opportunity to experience a time or a specific event that our ancestors experienced. It can be considered primarily as a leisure activity or a form of entertainment, which can lead one to question its authority in representing the past (Hunt, 2005). DeGroot (2008), however, examines the educational aspect when considering the immersive element of living history, finding that the reality of humans in living history far exceeds virtual or augmented reality often seen in museum exhibitions.

To conclude, WW2 and the 1940s persist in engaging the minds of the British people. There is an element of this time that people of today would like to keep and rejuvenate. Festivals and events inspired by the 1940s are a form of edutainment, which can provide an immersive or passive experience for attendees. This synergy between education and entertainment within 1940s-inspired festivals and events has the potential to provide a greater effective, engaging, and memorable experience. Discussions on authenticity often emphasise a bond with storytelling – keeping history alive, passing the true and real story down through generations, linking to personal and cultural identity and purpose, as well as nostalgia, a desire to connect with a more satisfying past, happy memories, or victorious times. Even if those memories are not ones of personal experience but are instead vicarious nostalgia, formed from a collective cultural memory.

Living history transports people back in time, where it may be experienced through re-enactment, which offers a simulation of the past and provides an entertaining educational experience. Festivals and events inspired by the 1940s have a prime opportunity to provide an authentic experience of the era, telling the story to create edutainment.

Methodology

This chapter, which is part of a wider study that concluded in 2021, aims to demonstrate the synergy between education and entertainment (edutainment) within 1940s-inspired festivals and events. Key areas of questioning include the importance of storytelling to disseminate memoirs of this historic era, how 1940s-inspired festivals and events can trigger nostalgia, the inclusion of edutainment and finding a synergy between entertainment and education, and the cultural duty to present an accurate representation of the past. There has been no previous comparison between what festival and event organisers believe they are providing in terms of edutainment and the attendee experiences. However, in this research, qualitative data was collected from both festival organisers and attendees to allow the researchers to interpret the opinions of both. Five themes were identified following data analysis:

1) Storytelling through living history to disseminate historic memoirs
2) Organisers of 1940s-inspired festivals and events feel a cultural duty to tell an accurate story of the era
3) Festivals and events inspired by the 1940s are a way to connect to the cultural and personal past
4) Triggers of nostalgia in 1940s-inspired festivals and events
5) The inclusion of edutainment in 1940s-inspired festivals and events

The data was collected during COVID-19 restrictions, using semi-structured online interviews via Zoom. This method enabled a flexibility for both the researchers to delve deeper into the interviewing process and the participants' ability to answer in depth. The questionnaire contained 30 open-ended questions linking to the main themes of edutainment, storytelling, living history, nostalgia, and cultural duty. An online ethics checklist was completed and an appropriate research strategy was developed to comply with the University's policies on integrity and professionalism for conducting ethical research. Promotional activities took place to generate attendee participants via LinkedIn and Facebook, as well as contacting key social media groups, such as *Love of the 40s* (2021) and *Events and Socials* Facebook pages. This generated ten participants familiar with 1940s-inspired festivals and events. Thirteen invitations were sent via Facebook Messenger and a further 21 email invitations were sent to 1940s-inspired festival and event organisers. A total of 34 UK-based 1940s-inspired festival and event organisers were contacted across the UK, which generated 13 festival and event organiser participants. The online interviews took place between 4 and 25 May 2021, with each interview lasting between 1 and 2.5 hours, which were recorded and transcribed for data analysis. In the research findings section below, attendee participants are referred to as 'A' and organisers as 'O' followed by their participant number.

Research findings

The inclusion of edutainment in 1940s events

Living history and re-enactment are both evident within 1940s-inspired festivals and events. They successfully present a form of edutainment for spectators and participants alike. The amalgamation of performance, authentic objects, information, and costume in 1940s-inspired festivals and events creates an atmospheric link to the past (Carneiro et al., 2019), aiding education (Jones, 2019), and providing a glimpse of what life was like (Chhabra et al., 2003). All research participants were asked to provide three words to describe the 1940s-inspired festivals and events they attended or organised. The results (see Figure 12.1) highlight the words mentioned more than once and show the synergy that is felt between education and entertainment. Specifically, the words 'educational', 'immersive', 'nostalgic', and 'fun' were repeated throughout the process.

The research identifies that festival and event organisers provide an edutainment experience; this is shown by the following quotes:

O6 – "the groups would bring items and talk about them in a relaxed way, from that sense it definitely did have an educational impact … we were able to educate people, it just wasn't so forceful; it was just keeping it relaxed and fun".

Figure 12.1 Attendee and organiser words to describe 1940s festivals and events.
Authors (2022).

O20 – "if it's not fun and entertaining, people will be bored, if it is fun they will learn, they will want to take part and come back".

O22 – "We want the educational side of things so we can get the children involved, it's trying to get a balance of the history, and the younger generation coming through the music and entertainment as well".

This research identifies that 1940s-inspired festivals and events can provide a form of education in an entertaining way (i.e. edutainment), providing there is engagement from both re-enactors and attendees. Organisers ensure that the re-enactors they welcome can speak with authority, research their subject well, providing authentic and historically accurate detail. The environment they create is as immersive and entertaining as possible, generating an authentic educational level of engagement in a fun and interactive way. This form of edutainment is deliberated further throughout the findings on the inclusion of storytelling, nostalgia, living history, and cultural duty within 1940s-inspired festivals and events.

Storytelling through living history to disseminate historic memoirs

The first theme identified from the research relates to the role of storytelling in 1940s-inspired festivals and events as an aid to disseminate historic memoirs. Storytelling is a powerful tool to connect the present with the past and educate future generations (The History Press, 2022). The role of storytelling is an integral part of 1940s-inspired festivals and events; this was clearly promoted by A19: "It's the sharing of stories, it's very much an oral history being passed down from people who were there and from people who have that connection". The application of storytelling within a festival and event setting facilitates an efficient exchange of information through a social interaction (Moscardo, 2020). Stories are often introduced to 1940s-inspired festivals and events through living history, an immersive historic narrative, bringing history alive, or an interpretation of a historic time (DeGroot, 2008). The staging of living history is supported by episodes of re-enactment. The purpose of which is to educate through performative means (DeGroot, 2008), educating the public through engagement and allowing them to develop an interest in and consume history.

The findings identified that education is a key factor in storytelling. Attendees are looking to be educated and to learn about the past to strengthen their historical cultural connection (Zhou et al., 2012). A17 stated, "We want to be educated about the past, but we want to learn from the past as well", and A3 confirmed, "It's how we understand how we all live now, by understanding how they lived in the past". Education is a key understanding of the meaning of storytelling and an element that is desired by attendees. A16 commented, "It is about the educational side of it, raising awareness of what went on, and how important that was to us as a nation". Similarly, the organisers strongly favour education through storytelling. This is substantiated by the organiser comments below:

> O18: "Everything we do is to make it as historically accurate as possible; we like to provide some form of educational elements within it and remembrance. We do tell the story and a lot of the local history".

> O8 stated, "You have got to be engaged with your audience. People do not want to approach you, so it is difficult to get your story over. I portray an American Paratrooper.... I cook my breakfast in front of everyone, you then get people coming over, the smell gets going, I talk to people about what I am cooking and how they cooked back then. You start the story small *(to get their attention)* and then build up and up".

> O13: "It is a more relaxed style of learning ... engaging and immersive, so you do not necessarily feel like you are being preached to or learning".

Attendees agree with this, reinforcing that storytelling is recognised and desired, A2: "It puts context around it *(the subject)* so that you can relate to a situation more". The participants agreed that education is an essential element within 1940s-inspired festivals and events. Authenticity through storytelling emanates from the re-enactors through a passion for their subject and a desire to share their knowledge. This supports Moscardo's (2020) efficient exchange of information. Providing a meaningful and memorable experience is evident in the following participant quotes:

> O15: "The amount of people that will come and say they have loved it and they have learnt so much. I think a big part of it is that you can go to school, read a book but there is nothing quite like going to an event and somebody going 'here hold this rifle its 80 years old and this is how it works', it's just that link to history that you can't get anywhere else".

> O14: "By the end of the weekend we feel as though we have been teaching someone at university, we never stop, it's so educational it's unbelievable".

Education is dependent on the engagement of both visitors and re-enactors; if they both engage, the holistic interaction can strengthen the educational experience (Knowles, 2016). Attendees acknowledge that they can learn through interacting with re-enactors:

> A19: These immersive events are more of a fun way to learn about history, if you go to a museum now most of them have a lot of hands-on events, which they have found is a better way to teach history because it comes alive. There is a big part of that in the 40s events, they are having fun, but they are learning at the same time.

> A21: They are able to educate in two ways, because the people coming along learn, but also the people doing it learn from the older visitors.

The channels to education and learning can be multi-directional, well-researched, and informed members of the re-enactment groups who may share their knowledge, allowing a tactile, immersive, educational experience for festival attendees. Whilst at the same time, they are constantly learning from other re-enactment groups, as well as visitors who are keen to share their own stories and knowledge of the time.

A cultural duty to tell an accurate story of the era

The organisers demonstrate a cultural duty to share the collective memories of the past, to remember, show respect, inform, and educate. O20: "We take this responsibility to make sure they are as accurate as possible very seriously", O13: The story must be "rounded and balanced", O9: "accurate and respectful". Organiser participants who actively embody the 'war years, both sides of the story' within their event, are careful to forewarn visitors of the German representation to allow those who may find the encounter unpleasant, the opportunity to opt out:

> O7: There was a fork in the path so if you didn't want to go through the German section you could opt out, but only one person chose not to go through.

> O18: They are not happy that they *(Germans)* are there, but they talk to them and by the end they come away realising they were doing a job; they didn't want to be there. A lot of these events will only show Britain at war, to me, you are deleting half of the story; there wouldn't have been a war without Germany, so therefore you have got to include them.

> O8: A lot of people don't think it's politically correct. You can't hide what happened, people need to know there was another side to the war, especially the children, we openly have Germans participating, they are just at one station. We want people to remember that the Germans were there, if you hide it, you are not portraying an accurate story.

Festivals and events inspired by the 1940s that highlight both sides of the story may shock attendees and challenge, for example, Noakes and Pattinson's (2014) theory of excluding the unsavoury elements of a cultural memory. It asserts that organisers feel the cultural duty to tell the story as accurately as possible, with strong educational benefits. The public are educated on O7 "the difference between the Nazis and the Germans", and it is apparent that this works, "by the end they *(visitors)* felt really sorry for them *(Germans)*, at the end of the day they were just everyday people".

Looking at the past in a hope to preserve or rejuvenate positive elements (Chhabra et al., 2003) may present a romantic, glorified reflection of the past (Hodge, 2012). Participants agreed that there may be an element of glorification. Moreover, they agreed it is acceptable to glorify a victorious and defining cultural moment if this is combined with a nuance of respect, remembrance, and pride, and not specifically a glorification of war:

> A3: "I don't see that there is anything wrong with that. It's a part of our past that happened and if no one is glorifying war, which I don't think they were, they are glorifying the positives. I think looking back at the lovely things of an era doesn't mean you are forgetting *(remembrance / respect)*".

> A21: "It glorifies but it's not a fault of them *(the events)*, if you are telling the truth it doesn't glorify, it enlightens".

> A16: "They are not glorifying war as such, but they are glorifying the period. It's back to that rose-tinted-specks again, you forget the pain and the suffering that went on".

The organisers endeavour to present an accurate story, and the research has evidenced that they feel a sense of cultural duty to do that to the best of their abilities. Whilst there is an element of glorification, exposing one to a more romantic reflection of the past (Hodge, 2012), this is not deemed to be negative in the re-telling of the actual story. The findings challenge Noakes and Pattinson's (2014) theory of cultural collective memories, as to fully exclude the less positive elements would not be considered an accurate re-telling of the story. Nevertheless, the organisers are mindful of the potential impact on attendees, and so it is managed with respect and balanced with desirable aspects of the decade that communities wish to preserve (Chhabra et al., 2003).

Connecting to the cultural and personal past

The social knowledge of the time has been passed from generation to generation through family stories (The History Press, 2022), keeping the history alive. This is corroborated by the following participant statements: A17: "We don't remember the war, we just understand from our grandparents", A2: "My mum talks about the war so that makes me interested", A21: "It makes people go home and ask about their own family, "you've got to know where you've come from before you can work out where you are going to". There is an essence of doubt that the sharing of cultural knowledge will continue in future generations, A3: "It was talked about through our family members but that will not happen for future generations", A2: "I don't talk to my kids, it will be interesting to see what happens over the next few years". The 1940s era feels easy to connect with:

> O22: "That relationship that is one or two generations away, so it's still relevant and exciting, there is still a connection".

> O12: "WW2 is the easiest comparison because there are still a few people around who remember it first-hand".

These generational links to the past (Noakes & Pattinson, 2014) create a personal remembrance:

> O20: "They absolutely love relating their family experiences to what they are seeing" ... even for the people who haven't lived through that era, they evoke a sense of the times that their parents or grandparents had".

> A19: "The German re-enactors only spoke German, they wouldn't break into English, it was the most peculiar experience, but to hear them in the uniforms speaking German, it just, event now I can feel goose bumps just thinking about it. My dad *(a D-Day Veteran)* never spoke about the war, he would talk about the comradery and his friends, but never spoke about the battles, it just gave me a real insight into how he must have felt".

Festivals and events inspired by the 1940s address the gap of cultural knowledge formed by those who could not talk about the hardships of WW2. An emotional engagement with history is stimulated through learning, respect, gratitude, and grief. This is paralleled with fun and a celebration of a victorious time, reflected in the music, fashion, and nostalgic tactile elements, a form of edutainment supporting the findings of Wagner (2015). This research has identified that 1940s-inspired festivals and events can aid a cultural and personal connection to the past. This is through a culmination of family and community stories remembered and shared during the experience. The connection is initiated by the festival/event, which once immersed into its holistic nature, memories are ignited. Family stories, combined with visual elements of living history, begin to come alive, providing a better understanding of what life was like and securing a cultural connection to the past.

Triggers of nostalgia

It is possible for both historical and personal nostalgia to be stimulated or triggered by 1940s-inspired festivals and events, with a focus on the holistic encounter generated through dress, music, entertainment, storytelling, re-enactment, and demonstrations. Whilst eight out of ten attendees felt a sense of nostalgia, they did not agree that it was a yearning for the past, as postulated by Chhabra (2012), that motivated their attendance. This is supported by A19:

> A19: "You get a lot of stalls selling things and you spend a lot of time saying 'oh, look we had one of those'. It brings back nice memories, but I certainly wouldn't want to be alive during the war, the camaraderie was lovely, but I wouldn't want to live in that time".

> A3: "When all these events are around the war, for me that takes away any wanting to look at it through rose tinted glasses. Patriotism, comradery, all that is great, but I think that's a positive side effect over something that people had no choice over. I wouldn't look back with nostalgia".

The 1940s was a decade filled with hardship, devastation, and destruction. To focus on the rose-tinted-glasses aspect of nostalgia suggests that cultures choose the more positive elements, a link with 'family' or a cultural connection to the past, for example, A16: "I think most people do that *(look back with rose-tinted-glasses)*, they see photographs and the people pulling together, community helping each other ... and you forget the years of pain and suffering...".

The organisers identify nostalgia with memories, remembrance, and reflection, focusing on the emotive element nostalgia may bring. However, this may not always be positive, as O9 commented,

> We are very careful that we don't go too much down the nostalgic route because it can be quite emotive for people, so it's not necessarily just the generation who lived through it, but the generations after that, who are going back through the memories.

The organisers agreed that nostalgia is important but hard to balance when you have the nicer and more positive elements of the music, fashion, comradery, and strength of a nation, contrasted with the hardships of war and its long-felt aftermath:

> O15: "Nostalgia is possible with things like the dance that we hold, the fashion pageant, that's all-nice nostalgia that you can recreate, the live music, even down to the pub that we have ... it's a caravan that turns into a 1940s pub with pictures of the king on the wall and the old pumps, things like that create nostalgia for people, but I don't think you can, and I don't think we would ever aim to say that we create a full nostalgic experience for people because it wasn't a great time".

Triggers of nostalgia at 1940s-inspired festivals and events have been identified by creating a word cloud (see Figure 12.2), which includes music, clothes, sounds, smells, and objects. These correspond with Cho et al. (2014) as the more common triggers sparking an idealistic recall of the past. A10 felt that "to dress in the 40s style with the 40s music, it's almost like you are taken back" in time. Today, one is exposed regularly to the music and fashion of the 1940s through films, TV, Radio, family photos, memorial events, and celebrations. Music was important in the 1940s to raise morale and evoke a sense of patriotism. The fashion was in some ways simple due to rationing, materials were in short supply, and everyone was encouraged to 'make do and mend,' fixing or handing down clothes, over purchasing new items (Hodge, 2012). Even so, the period created classic styles, the foundations of which still influence the fashion of today, a classic and elegant look that one can associate with.

The triggers identified in Figure 12.2, link to the senses and contribute to a holistic experience (Knowles, 2016), creating an element of immersive nostalgia through a simulation of the period, as festival and event organisers

Figure 12.2 Triggers of nostalgia identified by all participants.
Authors (2022).

confirm: O12: "The sights the sounds the smell, the costumes, the bunting, waving the union flags. The vehicles, you can still smell the leather from 80 years ago, and people still love all that. It's escapism for 24 hours".

This research confirms that 1940s-inspired festivals and events content can generate a positive nostalgic feeling, but attendees do not necessarily have a yearning for the past. The fun, warm, and positive features of nostalgia are evident, allowing a moment of escapism for those who experience a vicarious nostalgia. However, they are tinged with less positive images of war, and whilst organisers agree that it is important, and aim to generate an element of nostalgia for visitors, they are considerate that the event does not create a fully nostalgic experience.

Conclusion

Festivals and events inspired by the 1940s provide an emotional engagement with history, through the combination of multi-directional education, paralleled with fun and a celebration of a victorious time, which strengthens a cultural and personal connection to the past. This research confirms that 1940s-inspired festivals and events can generate a positive nostalgic feeling, reflected in the nostalgic tactile offerings. The event organisers identify nostalgia with memories, remembrance, and reflection, focusing on the emotive element of nostalgia. The event organisers agreed that nostalgia is important and can be triggered through clothes, music, sounds, smells, and objects. Organisers endeavour to present an accurate story, and the research has evidenced that they feel a sense of cultural duty to do that to the best of their abilities, often including the less positive images of the wartime, combined with an element of glorification, exposing one to a more romantic reflection of the past.

The research findings conclude that 1940s-inspired festivals and events can provide a form of education in an entertaining way, providing there is engagement from both re-enactors and attendees. The participants identified the key components of their 1940s-inspired festival or event experience as being educational, immersive, nostalgic, and fun. The festival or event must therefore create an environment that is as immersive and entertaining as possible, generating an authentic educational level of engagement in a fun and interactive way. The findings confirm that stories are often introduced to 1940s-inspired festivals and events through immersive living history, supported with episodes of re-enactment to educate through performative means. This engages attendees, who are looking to be educated, and allows them to consume history. Research participants agree that education is an essential element within 1940s-inspired festivals and events and attendees acknowledge that they can learn through interacting with re-enactors.

Declaration

This chapter draws on Zoe Leonard's previously unpublished MSc dissertation "A Critical Review of Authenticity, Storytelling and Nostalgia within 1940s events in the United Kingdom, to Determine if They Present an Authentic Reflection of the Past" – 2021.

References

Calder, A. (1992). *The People's War Britain 1939 – 1945* (Pimlico ed.). Pimlico.

Carneiro, M. J., Eusebio, C., Caldeira, A., & Santos, A. C. (2019). The influence of eventscape on emotions, satisfaction and loyalty: The case of re-enactment events. *International Journal of Hospitality Management, 82,* 112–124. Elsevier. https://doi.org/10.1016/j.ijhm.2019.03.025

Chhabra, D. (2012). Authenticity of the objectively authentic. *Annals of Tourism Research, 39,* 480–502. https://doi.org/10.1016/j.annals.2011.09.005

Chhabra, D., Healy, R., & Sills, E. (2003). Staged authenticity and heritage tourism. *Annals of Tourism Research, 30*(3), 702–719. https://doi.org/10.1016/S0160-7383(03)00044-6

Cho, H., Ramshaw, G., & Norman, W. C. (2014). A conceptual model for nostalgia in the context of sport tourism. *Journal of Sport & Tourism, 19* (2), 145–167. https://doi.org/10.1080/14775085.2015.1033444

DeGroot, J. (2008). *Consuming history: Historians and heritage in contemporary popular culture* [ProQuest Ebook Central] (2nd ed.). Routledge.

Fog, K., Budtz, C., & Yakaboylu, B. (2005). *Storytelling: Branding in practice* (1st ed.). Springer Verlag Berlin.

Gao, J., Lin, S., & Zhang, C. (2020). Authenticity, involvement and nostalgia: Understanding visitor satisfaction with an adaptive reuse heritage site in urban China. *Journal of Destination Marketing and Management, 15,* 100404. https://doi.org/10.1016/j.jdmm.2019.100404

Getz, D., & Page, S. J. (2016). Progress and prospects for event tourism research. *Tourism Management, 52,* 593–631. https://doi.org/10.1016/j.tourman.2015.03.007

González-González, J.-M., Jesús-Gerardo, F.-C., & Darío, E.-S. (2022). Educating in history: Thinking historically through historical re-enactment. *Social Sciences, 11*, 256. https://doi.org/10.3390/socsci11060256

Goulding, C. (2002). An exploratory study of age-related vicarious nostalgia and aesthetic consumption. *Advances in Consumer Research, 29*, 542–546.

Hill, J. (1999). Postscript: A war imagined. In N. Hayes & J. Hill (Eds.), *Millions like us? British culture in the Second world war*. Liverpool University Press.

Hodge, S. (2012). *The home front in World War Two: Keep calm and carry on*. Pen and Sword Books Ltd.

Huber, J., Caine, V., Huber, M., & Steeves, P. (2013). Narrative inquiry as pedagogy in education: The extraordinary potential of living, telling, retelling, and reliving stories of experience. *Review of Research in Education, 37*, 212–242. https://doi.org/10.3102/0091732X12458885

Hunt, S. J. (2005). Acting the part: Living history as a serious leisure pursuit. *Leisure Studies, 23*, 387–403. https://doi.org/10.1080/0261436042000231664

Imperial War Museum. (2022). *10 photos of VE Day celebrations* [online]. London. https://www.iwm.org.uk/history/10-photos-of-ve-day-celebrations

Jones, K. (2019). *The destination for history – The History Press – The art of historical storytelling* [online]. Cheltenham. https://www.thehistorypress.co.uk/articles/the-art-of-historical-storytelling/

Kirka, D. (2022). *Stories of UK's disappearing World War II generation* [online]. The Associated Press. https://apnews.com/article/queen-elizabeth-ii-veterans-england-world-war-c8ad03cf77048c812894442ad560fabc

Knowles, B. (2016). *Re-enacting the second world war: History, memory and the UK Homefront* [online]. Thesis (PhD). University of Manchester. https://www.semanticscholar.org/paper/Re-enacting-the-Second-World-War-%3A-history%2C-memory-Knowles/a497a2caea4a8f7f7da2bce3a9938b5b4aa43711

Last, N. (1939). The second world war diaries of housewife, 49. In R. Broad & S. Fleming (Eds.).(2006), *Nella Last's war: The second world war diaries of housewife* (2nd ed., Vol. 49, p. 184). Profile Books Ltd.

Love of the 40s. (2021). *Events* [online]. Rustic Bridge. https://loveofthe40s.co.uk/

Moscardo, G. (2020). Stories and design in tourism. *Annals of Tourism Research, 83*, 102950. Science Direct, Elsevier, Stories and Design in Tourism. https://doi.org/10.1016/j.annals.2020.102950

Noakes, L. (2014). War on the web: The BBC's 'People's War' website and memories of fear in wartime in 21st-century Britain. In L. Noakes & J. Pattinson (Eds.), *British cultural memory and the second world war* (pp. 47–65). Bloomsbury Publishing Plc.

Noakes, L., & Pattinson, J. (2014). The cultural memory of the second world war in Britain. In L. Noakes, & J. Pattinson (Eds.), *British cultural memory and the second world war* (pp. 1–24). Bloomsbury Publishing Plc.

Packer, J., & Ballantyne, R. (2004). Is educational leisure a contradiction in terms? Exploring the synergy of education and entertainment. *Annals of Leisure Research, 7*(1), 54–71. https://doi.org/10.1080/11745398.2004.10600939

Prayag, G., & Del Chiappa, G. (2021). Nostalgic feelings: Motivation, positive and negative emotions, and authenticity at heritage sites. *Journal of Heritage Tourism* [online], (22 January 2021). https://doi.org/10.1080/1743873X.2021.1874000

Railway in Wartime Weekend. (2022). *Whitby Events* [online]. Whitby. https://www.whitbyevents.co.uk/index.php?com=detail&eID=1065

Rodriguez, A. R. (2016). *Re-enactment vs. living history…and the renaissance faire? Heroes, heroines and history* [online]. USA. https://www.hhhistory.com/2016/11/reenactment-vs-living-historyand.html

Santonen, T., & Faber, E. (2015). Towards a comprehensive framework to analyse edutainment applications. In *Shaping the frontiers of innovation management. Proceedings of the XXVI ISPIM conference*, Budapest, Hungary. ISBN 978-952-265-779-4.

The History Press. (2022). *The destination for history – The History Press – Storytelling* [online]. Cheltenham. https://www.thehistorypress.co.uk/localhistory/storytelling/

Thompson, K. (2019). *The Stepney Doorstep Society: The remarkable true story of the women who ruled the East End through war and peace* [Kindle]. Penguin.

Wagner, B. (2015). The seriousness and fun, when edutainment is associated with history teaching. In *International Journal of Research on History Didactics, History Education, and History Culture* (JHEC). P155 – 165 in 36/2014.

Wyatt, B. (2022). Edutainment. In D. Buhalis (Ed.), *Encyclopaedia of tourism management and marketing*. Edward Elgar Publishing. https://doi.org/10.4337/9781800377486.edutainment

Wyatt, B., Leask, A., & Barron, P. (2021). Designing dark tourism experiences an exploration of edutainment interpretation at lighter dark visitor attractions. *Journal of Heritage Tourism*, *16*(4), 433–449. https://doi.org/10.1080/1743873X.2020.1858087

Zhou, X., Wildschut, T., Sedikides, C., Shi, K., & Feng, C. (2012). Nostalgia: The gift that keeps on giving. *Journal of Consumer Research*, *39*, 39–50. https://doi.org/10.1086/662199

13 Playing with Your Food

Fun, Frolics, and Learning at Food Festivals

Ingrid Kajzer Mitchell, Christine Van Winkle and Will Low

Introduction

This chapter explores how an edutainment framework where education and entertainment occur simultaneously moves the food festival experience beyond didactic processes to fun and frolics that result in experiential and deep learning. As social phenomena encountered in virtually all human cultures (Falassi, 1987), festivals constitute important research sites for understanding how and what people learn whilst engaging in tourism (Falk et al., 2012; Rossetti & Quinn, 2019; Van Winkle & Lagay, 2012). Food festivals, in particular, are distinctive within the growing spectrum of festivals (Getz, 2010; Wilson et al., 2016) in the ways they can engage all our senses (Chang, 2011; Organ et al., 2015; Richards, 2021), opening up even more opportunities for immersive and interactive learning to occur through playful interactions. The authors use an inductive multiple case-study research design, drawing on data from the Local Wild Food Challenge (LWFC) and the Harvest Moon Festival (HMF) to examine how these two food festivals engaged their audiences in deep learning processes by inviting people to 'play'. This approach reflects a constructivist philosophy and desire to gain insight into the experiences of participants (Adom et al., 2016) from their own perspectives.

Learning experiences at festivals have been less commonly studied than physical, cultural, social, economic, and psychological aspects (Getz, 2012; Van Winkle & Bueddefeld, 2016). This is even more the case with food festivals. However, learning is often a motive for and an outcome of attending a festival (Crompton & Mckay, 1997; Tölkes & Butzmann, 2018). Learning through entertainment at a food festival can be better understood if the festival is considered a ludic learning space (Kolb & Kolb, 2010), which is a space 'relating to, or characterised by play' (Merriam Webster, n.d.). Merging fun and learning has a long history as a pedagogical approach, from the mid-1800s (Froebel's kindergarten movement) to the late 1800s (Maria Montessori's philosophy of education), and on to the early 1900s (Piaget's 'stages of cognitive development'). These formulations of 'learning as play', or 'learning through play', may have focused on understanding childhood development (Smith & Pellegrini, 2008), but "play is a strategy for learning at any age" (Tatter, 2019, p. 10).

DOI: 10.4324/9781003305415-17

Indeed, according to Kolb and Kolb (2010), referring to work by Huizinga in 1950, humans should be considered *Homo Ludens*, literally the man (person) who plays, as much as *Homo sapiens*, the man (person) who knows.

How learning occurs through this play can be understood by applying the ideas of co-creation proposed by Prahalad and Ramaswamy (2004) and service ecosystems by Vargo and Lusch (2016). These ideas are central to research into festivals that increasingly focuses on how attendees play a lead role (Richards, 2015), actively contributing to, and therefore co-creating event experiences (Rachão et al., 2020) where entertainment and learning coalesce (Rossetti & Quinn, 2019). The role of the festival attendee changes from a person in the audience who passively receives an experience, to being an actor actively engaged in creating value by integrating resources within a broader network of festival stakeholders (Richards, 2015). This network of actors, resources, and institutional arrangements is called a service ecosystem (Vargo & Lusch, 2016; Wieland et al., 2012). Within the festival service ecosystem, attendees learn by playing with each other and with their food.

Thus, by exploring ludic learning spaces and service ecosystems within the context of food festivals, this chapter suggests how edutainment in a food festival context can lead to learning *about* food and *through* food. The food festival is conceived of as a service ecosystem and a ludic learning space in which interactions between multiple actors result in multi-layered edutainment experiences. Analysis of the data gathered at the LWFC and HMF considers various levels of interaction, from individual to relational, and from personal to communal, to reflect on festival goers' motivations for attending, what was learned, and how learning occurred. The findings reveal how the entertainment and learning that occurs at a food festival are dependent on the willingness of actors in the festival setting to interact and share resources – a central aspect of service ecosystem thinking. The findings also reveal how the ludic learning space is elastic and how festival experiences do not just constitute a 'fun day out' but that the fun and frolics encourage attendees to explore their creativity, express themselves in novel ways, and learn about alternative food realities beyond the time and space of the festival itself.

Food festivals and edutainment experiences

Food festivals bring a broad range of actors and resources together in a place-based context (Lau & Li, 2019), merging entertainment and learning (edutainment) by connecting people through their shared knowledge and skills (or desire to learn skills) about food and through their shared consumption of foods to co-create festival experiences (Organ et al., 2015; Richards, 2015). The temporal and physical space of a festival can be viewed within an edutainment framework as a ludic learning space, which Kolb and Kolb (2010) define as having three characteristics: it is free, it entails stepping out of 'real' life, and it is bounded in space and time. The three conditions are not independent of each other, as Edensor (2014) observes: "Festivals break up the usual routines

and rhythms of everyday life, marking out a temporary departure from …
usually unreflexive practices during which ordinary conventions may be sus-
pended and participants allowed greater expressive, creative and improvisa-
tional licence" (p. 3).

Edutainment also moves the festival experience beyond an exchange
between producers and consumers where value is created when a tangible good
is sold to a passive consumer. Edutainment shifts the focus to the 'experience
economy' (Prahalad & Ramaswamy, 2004) where attendees co-create and consume 'fun' experiences. Expanding on the concept of co-creation proposed by
Prahalad and Ramaswamy (2004), authors such as Pinho et al. (2014), Lusch
and Vargo (2014), and Lusch and Nambisan (2015) characterise co-creation
occurring in a service ecosystem whereby "resource-integrating actors" are
"connected by shared institutional arrangements and mutual value creation
through service exchange" (Vargo & Lusch, 2016, pp. 10–11). Within the context of food festivals, scholars have noted a range of levels of engagement
amongst festival actors, from passive observants to active participants (Organ
et al., 2015; Rachão et al., 2021), which in turn may impact the degree of
co-creation that occurs in the edutainment experience.

Central to viewing food festivals through a service ecosystem lens is the
notion of interactions being the driving force in the co-creation process
(Fyrberg & Jüriado, 2009). Each festival actor or stakeholder becomes engaged
in "non-everyday activities of self-expression and interaction with other individuals" (Lefrid & Torres, 2022, p. 369) and brings resources – unique skills,
knowledge, and experience – to create value (Vargo & Lusch, 2011, 2016). The
level of interaction at food festivals is typically very high, where food providers
and consumers are able to engage in knowledge sharing and discuss the origin
of the food and the food production process (Organ et al., 2015).

The social interactions people engage in at festivals and visitor attractions
are critical to the process of learning (Axelson & Swan, 2010; Packer & Ballantyne, 2005). The free-choice learning literature describes the interacting role of
the personal, socio-cultural, and physical contexts within the contextual model
of learning (Falk & Dierking, 2000). The personal context includes the personal history one brings to a learning situation and includes past experience,
knowledge, motivation, and expectations (Falk & Storksdieck, 2005). These
interacting contexts will influence how and what festival participants learn
when learning is not compulsory.

The informal learning experiences (Heim & Holt, 2022; Paris, 1997) and
'learning for fun' occurring at food festivals, which transcend the ordinary
(Falassi, 1987), contribute to potentially transformative experiences (Neuhofer
et al., 2020). Participants' direct involvement with food, whether through har-
vesting and preparing foods, participating in cooking demonstrations, and
competitions and sampling food (Getz & Robinson, 2014) may break habitual
behaviour and elicit changes in consumption practices (Organ et al., 2015). The
recognition of food as a medium for deep learning is well recognised and more
extensively studied outside of festival scholarly work. Adult education and

food studies scholars have identified food as both an "object" in itself to learn about and as "a vehicle for learning" (Flowers & Swan, 2012, p. 423). Food as a vehicle for learning can be particularly transformative in two regards. First, where food comes from and what constitutes food draws us back to nature and centres the non-human world (Kajzer Mitchell et al., 2017; Levkoe, 2006). Learning about food can be an "opening *[to]* the possibility of more inclusive, and more sustainable, ways of life" (Sumner, 2008, p. 352), such as increasing the role of local food in a particular region more broadly (Fontefrancesco & Zocchi, 2020). Second, preparing and consuming food is a highly social activity rather than just about individual need for nutrition, centring family, and social relationships (Bourque et al., 2020).

Local Wild Food Challenge and the Harvest Moon Festival – A dual case study

The two festival case studies that this chapter explores – the Local Wild Food Challenge (LWFC) and the Harvest Moon Festival (HMF) – were each part of separate larger research projects exploring the learning experiences of people attending festivals during 2015 and 2017, respectively. The LWFC is a one-day festival that takes place annually in communities across the world (e.g. United States, Italy, Finland, and New Zealand). The challenge aspect of the LWFC is a relaxed and enjoyable cooking contest which sees young and old, amateur and professional cooks, and foragers coming together with volunteers and sponsors to showcase "the resourcefulness of local people living in communities where wild food is available" (Local Wild Food Challenge, n.d.). The HMF is a multi-day annual festival held in a small rural Manitoba town that "celebrates the harvest season and local food production"; it may even be considered a hybrid music-food festival with its heavy emphasis on music over the weekend event Harvest Moon Festival, n.d. 'What is Harvest Moon'). Both events are held outdoors and include various co-creation opportunities through cooking competitions, food demonstrations, hands-on activities and workshops, presentations by local food experts, food-producer stalls, and music, all centred around the theme of enjoying and learning about alternative foods such as local sustainable food production and wild food foraging.

The two festivals can be positioned as ludic spaces "that are occasions for pleasure and fun" (Edensor, 2014, p. 3). Both create immersive, multi-sensory, playful experiences where one might be sampling wild racoon meat with foraged herbs (LWFC) or socialising by enjoying prairie music while eating locally grown food (HMF). These two festivals exhibit the first characteristic of ludic spaces discussed in the previous section by utilising an open space and format, providing entertainment and education opportunities through free-choice, informal, non-compulsory learning (Heim & Holt, 2022; Paris, 1997), where festival attendees have a high level of autonomy to choose what, where, how, and with whom to learn (Falk & Dierking, 2000). The second characteristic is reflected in festivals being special occasions where one is absent from work or the everyday, to "celebrate life in its 'time out of time'" (Falassi, 1987, p. 7). Finally, the LWFC and the HMF are bounded in space and time, occurring on a designated physical site over one or two days, though crucially, we argue that the edutainment experiences are not just occurring in situ but emerge before, during, and after the experience (Antón et al., 2018; Ind et al., 2013; Wyatt et al., 2021). In addition, the festivals operate as service ecosystems, which emphasises the 'how' of learning in these ludic spaces (Axelson & Swan, 2010; Grenier, 2010; Kanhadilok & Watts, 2014), where the social interactions people engage in at festivals co-create fun and learning experiences.

Methods

Data collection for both case studies was approved by Research Ethics Boards at Royal Roads University and University of Manitoba, respectively. Data were primarily collected through 60 on-site semi-structured interviews (30 at LWFC in Martha's Vineyard and 30 at HMF). The principal researcher or a research assistant approached attendees at random and if a person indicated an interest in being involved, they would take part in guided interviews lasting approximately 10–20 minutes. For both the LWFC and HMF, a protocol of open-ended questions allowed prompts for further questions when needed (Bryman, 2012). The questions explored the reasons why people participated in the two respective events, their prior experiences and understanding of alternative foods, and what participants felt was significant about their experiences, including if and how they felt the experience had changed how they thought about foods. In the case of the LWFC, additional data sources included supplementary off-site interviews with festival participants and the founder/ organiser for further insights and clarification. These off-site interviews took place virtually over the phone or via a virtual meeting software, such as BlueJeans and Zoom, either prior to or after a LWFC.

Due to the immersive nature of the food festival environment (Neuhofer et al., 2020), the researchers also observed elements of each festival setting (e.g. watching participants engage in cooking, spontaneous conversations with others, and teaching each other new skills), and interacted through participation (e.g. sampling and trying new food). This "insider/outsider" experience is a defining element of more participatory observational research (Jaimangal-Jones, 2014). At times throughout the day, the researcher gathered observations as an insider, actively engaging in edutainment activities, and at other instances the researcher observed as an outsider, watching the behaviours and emotions of others in the festival setting. At the end of the day, the researchers captured additional field notes of actions and interactions observed. Combined, these observations provided additional context, which Kearns (2000) and Mackellar (2013) suggest helps to garner complementary insight into the relationships and interactions between audience groups, as well as the relationships between the festival activities and the different attendees.

Table 13.1 Motives for attending the food festivals

	Local Wild Food Challenge	Harvest Moon Festival
"Fun" Motives	"something to do" "it's a beautiful day", "it is a really nice day out" "[it is a] fun experience"	"listen to music, decompress and enjoy" "fun day outside with kids"
Novelty/Curiosity Motives – **trying new things**	"to try some new food" "not what you normally expect" "It captured my imagination"	looking for a "new experience" "curious about the festival, never been before"
Escape Motives – **break from daily activities, escape from the usual environment**	"get outside"	"a break from the city"
Sociability Motives – **interaction with others**	"meeting friends" "helping a friend" "a buddy is cooking a pig" "friends with volunteers" and "organiser"	"time with family" "see friends"
Communitarian Motives – **being a part of the "festival community"**	"supporting local" "important community event"	"sharing traditions"
Learning Motives – **learning skills and gaining knowledge**	"broaden my horizon" 'interested in what other people are foraging'	"learn about food"

Open-ended interview questions and observations were analysed through an iterative qualitative content analysis to describe findings. Assarroudi et al. (2018) suggest this systematic categorisation enables researchers to explore the interview transcripts and field notes to uncover trends and patterns. Analysis thus began by exploring the whole and then proceeded with open coding, where categories were created and similar ideas or statements were grouped together. This resulted in categories and subcategories describing the topic of interest, such as motives for attending, what was learned, and how learning occurs (see Table 13.1).

Results and discussion

In keeping with the aims of this book, the conceptual framework that merges ludic learning spaces and service ecosystems is used to explore how people co-create edutaining food festival experiences and their perceptions of the resultant personal learning outcomes. The processes through which learning and entertainment merge occur at different levels of aggregation, from individual to relational depending on the types of resource integration involved. Similarly, the outcomes of edutainment have impacts that can be personal or they can have communal implications.

Whilst learning was often cited at both events amongst their reasons for visiting, not everyone attending the festivals was intrinsically motivated to learn. Many of the motivations to attend were more instrumental, especially to have a 'fun day outside' with family and friends. Alongside 'fun' and 'learning', the other broad categories of motives for attending the event revealed by the analysis are novelty, escape, sociability, and communitarianism (see Table 13.1).

The motives participants expressed for attending the two food festivals are consistent with those identified previously by food festival scholars, among others, escaping a daily routine, creating memorable experiences, and spending time with others (Chang, 2011; Horng et al., 2013; Lefrid & Torres, 2022). They also align with other festival scholarly literature (cf. Crompton & McKay, 1997), most notably Tölkes and Butzmann (2018), whose study of a Munich music festival identified how a desire to learn about more sustainable lifestyles can co-exist with entertainment motives of 'fun' and 'enjoyment'. Thus, HMF participants not only engaged in a range of workshops where they played and experimented with aspects of sustainable lifestyles but also relaxed and enjoyed live music. Similarly, LWFC festival goers attended demonstrations to learn new cooking techniques for wild meats and foraged plants but also immersed themselves in the 'playful' tasting of these foods (e.g. using their hands or sticks to sample food). This in turn created a memorable eating experience by providing participants with an alternative perspective on the food object (Spence, 2022). As one attendee remarked, "Who knew you could eat racoon meat and pickled worms are tasty!" Echoing Heim and Holt (2022), these self-directed learning experiences are both personally interesting and enjoyable.

The motivation to learn at the food festivals aligns with Paris' (1997) findings (in the context of museums rather than a food festival) that free-choice learning emphasises autonomy to develop one's own learning agendas and goals, reflecting what Kolb and Kolb (2010, p. 47) refer to as creating one's own "game rules and conduct", and thereby contributing to more authentic and higher order learning, or deep learning. The LWFC attendees who choose to participate in the food festival competition are guided by their personal interest in deciding what to forage, how to prepare for the event's competition, and how to assemble dishes, and are thus taking charge of the learning process that unfolds.

What was learned

When describing their learning experiences, many attendees at both events spoke about *learning to appreciate*. LWFC attendees described learning to appreciate alternative foods, or "to think more consciously" about different foods by tasting them. For instance, one participant noted: "I tasted new things, I'm not sure about the wild goat ... but I appreciate different techniques to prepare food". At HMF, attendees came to appreciate through the whole festival experience how their own actions contributed to the broader food system and, therefore, their role in sustainable food production and consumption.

One HMF attendee highlighted "caring for each other and trying to eliminate waste" as part of what s/he had learned to appreciate. Another noted gaining a greater appreciation and "respect for the growth and effort behind food".

Attendees at these two festivals also talked about *learning specific skills and knowledge*; they often described specific food preparation techniques. At the LWFC, attendees learned about foraging and food preparation skills; one participant described: "Learning about deer preparation. If you drain the blood sooner, it gets tastier". At the HMF, attendees talked about learning what local food producers had to offer and how to prepare these items, as well as "what I can do to support local *[food production]*".

Other learning outcomes included *challenging habitual attitudes and beliefs* towards alternative food sources; Flowers and Swan's (2012, 2015) work suggests a "food pedagogy" is at play where the attendees are challenging normalised food stereotypes. For instance, LWFC attendees remarked, "That wild food is more readily available", or that local produce is more "inexpensive" than previously thought, and that certain foods previously perceived to be inedible can be consumed (e.g. "You can actually eat racoon"). The LWFC further emphasises demystifying alternative foods. In the words of the LWFC organiser and founder:

> We've educated people over the last decade as to … what's possible … some of the foraged plants have been things that you walk past every day but you don't realise that they're pretty good eating … some fish species that people thought they'd never want to use … things people throw away … we'll debunk that myth that this is just crap fish, it is not.

How learning occurred

Research participants at both events discussed formal and informal learning opportunities, shaped by various acts of co-creation (Vargo & Lusch, 2011). Formal learning opportunities at the LWFC included the competition (participating, actively experimenting) and demonstrations (observing a chicken being plucked or deer being skinned). At the HMF, participants talked about the workshops they attended on diverse topics such as mycelium (fungus) and beekeeping, as well as the demonstrations about food production techniques. Informal opportunities included descriptions of playing "with ideas in the creative process" (Kolb & Kolb, 2010, p. 27) of food preparation. As described by one LWFC attendee: "It is quite cool seeing different ways people use the same ingredient that I have [...] sometimes they do it differently and sometimes I learn stuff".

As Van Winkle and Lagay (2012) explain, learning at a food festival occurs through multi-sensory means, such as taste and smell (e.g. tasting new flavours or smelling racoon meat) or through direct contact with food, whereby attendees can feel novel textures in their hands or mouth. Such findings were also revealed within this study. Learning also occurred through observing and seeing, for example, by watching others try novel foods, or through visual exposure to new ways of presenting food. Listening to others tell stories was another important learning strategy and contributed to an awareness and a willingness to try new things. Following a service-dominant logic, casual, self-directed impromptu conversations between attendees encouraged an exchange of specialised skills and knowledge, which corresponds with Vargo and Lusch's (2011) findings. For example, where and how to forage and how to prepare food in new ways. As noted by an LWFC festival attendee: "I struggled for a long time to come up with a good way of cooking rabbit [...] I was talking to a chef and we were talking about different ways". In the LWFC organiser's own words: "Ideas get exchanged a lot during the day's activities, and these skill sets, ideas and sensibilities get shared".

Multi-layered learning

The responses discussed above highlight and contrast how learning occurs not only at an individual level but also collectively, as festival ecosystem actors learn from each other and together, with the learning processes and outcomes shaped by the nature of a ludic space and the playful interactivity inherent in this service ecosystem.

Deliberate institutional arrangements at the LWFC (e.g. the cooking competition) and the HMF (e.g. experiential demonstrations) facilitate interaction, resource integration, and encourage individuals to develop their creative potential and capacity to innovate (Richards & Raymon, 2000). Thus, as revealed in Lusch and Vargo's (2014) study, the food festival structures are 'making room for co-creation'; individual actors actively integrate existing and new resources (food skills, knowledge, and competencies), listen, provide, and take feedback, and subsequently adapt and recombine them to create more resources.

Personal level

What and how attendees learn is exemplified by individual actors playing a key role in the process of co-creating experiences (Lusch & Vargo, 2014), whether through active experimentation, interaction, and manipulation of objects in food workshops, or through "play[ing] with their potentials" (Kolb & Kolb, 2010, p. 27) during the competition. The LWFC contestants displayed their personal competency by mobilising their own knowledge and skills, and their creativity by bringing to life new menus, novel recipes, and food presentation. Hands-on food experimentation provided the HMF attendees with learning opportunities that called upon innovation and prior experience and provided potential for immediate feedback. This form of immersive play is founded on the notion that learning and understanding is possible when individuals can manipulate and interact with objects and event exhibits (cf. Grenier, 2010).

This individual participation led many attendees to report learning that was very personal, contributing to their personal growth, and development of their understanding of food outside of the ludic festival space. Attendees spoke of personal empowerment and gaining the courage to try to more actively source their own food. For example, an LWFC attendee shared:

> It gives you a little window into the fact that I don't have to go to the shop and buy everything, there's lot of great ways to sustain yourself, including foraging and gleaning in the fields, and that's kind of something that's nice that I'm learning.

Interactional

Social interaction with others is identified by Axelson and Swan (2010) as an important part of how learning occurs at events. Resource sharing and recombination of resources amongst attendees is supported through social interaction and collaboration among a wide ecosystem of actors such as local food experts and family members. How we learn at edutainment festivals is also influenced by the social nature of the experiences themselves (Packer & Ballantyne, 2005). The LWFC and HMF include enjoyable and entertaining activities, such as music, tasting, cooking demonstrations, and what Kanhadilok and Watts (2014, p. 27) refer to as "play-zones", where socialising and spontaneous participation and interactions encourage people to be imaginative, and attendees take on playful identities. From a service ecosystem perspective, the LWFC competition and the HMF festival program become part of a broader "architecture of participation" which Lusch and Nambisan (2015) suggest animates the edutainment experiences whilst coordinating, guiding, and supporting the interactions and resource exchange needed amongst actors in the ecosystem.

While interactions between festival attendees, and the social nature of the experience itself, play a key role in the edutainment festival experience, the ludic learning space is also co-created through human-to-nonhuman interactions. It is not only the humans who, through play, are producing and sharing knowledge (Kajzer Mitchell et al., 2017). The non-human natural elements such as plants and soil also become catalysts for creativity, and vehicles for learning (Flowers & Swan, 2012), both informing and mobilising our understanding of foods. For example, as attendees are interacting with food-related activities, food displays, and plated food, they are connecting not only with their personal values and deepest desires but also with the ecological system and are being taught about the world we live in. In other words, the various food practices (gathering, foraging, etc.) and food outcomes (new recipes, plated dishes) become performative resources through which people learn (Flowers & Swan, 2012).

Relational and communal

At the relational and communal levels of analysis, the food festival experience is enhanced by the authentic and reciprocal relationships (Organ et al., 2015) between festival organisers, the attendees, and the festival community. At both festivals, many attendees are returning participants and come to know each other by name, and this sustained relationship over time allows for learning to evolve from one festival experience to another. The recursive nature of the LWFC and the HMF, with people attending the event multiple times, Kolb and Kolb (2010) suggest, gives continuity for the individual's experience to mature and deepen, moment to moment, event to event, year to year. As shared by one LWFC participant:

> I definitely recognise people that have done it the year before, like there's people that do it regularly. And some of the people that help out on the day, they quite often come over and talk to me while I'm cooking, that's nice, and they remember my name and stuff.

The research analysis also illustrates how ecosystem actors who have an ongoing relationship, such as families, use free-choice edutainment events as tools or resources to build a family identity, and a common history and narrative (Bourque et al., 2020) beyond the confines of the festival setting. For example, in the case of LWFC, attendees shared stories of how they, as a family, think about and discuss the festival 'all year around'. Thus, the edutaining festival experience is not only an important resource integration site on the day of the event but becomes a "transitional space" (Kolb & Kolb, 2010, p. 31) where the co-created explorations extend beyond the day itself and continue to provide opportunities to create meaningful experiences even after attendees leave (Wyatt et al., 2021). Deep learning is enabled as festival attendees may "come back to the familiar experience with a fresh perspective" (Kolb & Kolb, 2010, p. 47) and "develop new goals and pursue new opportunities" (Lusch & Nambisan, 2015, p. 169) as actor interactions unfold over time (cf. Ind et al., 2013). As one LWFC returning attendee remarked when describing their entry to the competition element of the festival: "I'm used to just buying stuff, but now I can go out foraging with [another LWFC actor] and we figure out what mushrooms we're going to use". Similarly, attendees at the HMF shared how their festival experiences led them to consider "methods for food production in my own yard" and seeking out and making "healthier decisions and choices".

Conclusion

Research on food-related learning processes in adult education and in food studies positions food not only as an object of learning but also as "a vehicle for learning" (Flowers & Swan, 2012, p. 423). Sumner (2008, p. 356) was probably not intending a double entendre when stating that food can act as "an entree into larger questions about how we live, how we relate to each other and how we relate to earth", but this chapter has shown how the edutainment festival experience can lead to learning *about* food and *through* food. The learning that some attendees reported goes beyond information transfer and beyond the time and space of the festival. If participants begin to (re-)examine how "they cook, shop, and eat" (Flowers & Swan, 2015, p. 8), this may catalyse "a transformative festival learning experience (Neuhofer et al., 2020), opening the possibility of more inclusive, and more sustainable, ways of life" (Sumner, 2008, p. 352).

While this chapter has used a conceptual framework merging service ecosystems and ludic learning spaces to understand the interviews and observations at two food festivals in the contexts of edutainment, the goal was not to quantify instances of co-creation or categorise individual festival events as edutainment. Instead, this chapter suggests that in these two festivals, the edutainment experience that emerges is dependent on actors' own motivations, actions, and interactions, and the actions of other actors – a central aspect of service ecosystem thinking (Pinho et al., 2014). Further, the experience is not bounded by time and space of the festivals; the ludic learning space is more elastic, stretching out to pre- and post-learning processes consistent with Kolb's (1984) experiential learning cycle (experiencing, reflecting, thinking, and acting), which may be deepened when festival goers return to the festival year after year.

In terms of future steps to understanding the learning properties and dimensions of edutainment festival experiences and activities, more research is needed on the types of learners or attendee segments, actor interactions (human and non-human), and co-created outcomes between a broader set of festival stakeholders as they evolve over time. Building on the work of Organ et al. (2015), it is important to examine the edutainment festival experience as a long-term process where value creation starts before and lasts long after attending the festival. Moreover, the service ecosystem literature argues that the quality of interactions is essential to value creation (cf. Fyrberg & Jüriado, 2009) and future festival research would benefit from better understanding of how factors such as trust, power, and competition may impact how attendees actively participate in and consume edutainment festival experiences.

References

"About" (n.d.). The local wild food challenge. https://www.localwildfoodchallenge.com/about/

Adom, D., Yeboah, A., & Ankrah, A. K. (2016). Constructivism philosophical paradigm: Implication for research, teaching and learning. *Global Journal of Arts Humanities and Social Sciences*, 4(10), 1–9.

Antón, C., Camarero, C., & Garrido, M. J. (2018). Exploring the experience value of museum visitors as a co-creation process. *Current Issues in Tourism*, 21(12), 1406–1425. https://doi.org/10.1080/13683500.2017.1373753

Assarroudi, A., Heshmati Nabavi, F., Armat, M. R., Ebadi, A., & Vaismoradi, M. (2018). Directed qualitative content analysis: The description and elaboration of its underpinning methods and data analysis process. *Journal of Research in Nursing*, 23(1), 42–55. https://doi.org/10.1177/1744987117774166

Axelson, M., & Swan, T. (2010). Designing festival experiences to influence visitor perceptions: The case of a wine and food festival. *Journal of Travel Research*, 49(4), 436–450. https://doi.org/10.1177/00472875093467

Bourque, C., Houseal, A. K., & Welsh, K. M. (2020). Free-choice family learning: A literature review for the national park service. *Journal of Interpretation Research, 19*(1), 7–29. https://doi.org/10.1177/109258721401900102

Bryman, A. (2012). *Social research methods* (4th ed.). Oxford University Press.

Chang, W. (2011). A taste of tourism: Visitors' motivations to attend a food festival. *Event Management, 15*(2), 151–161. https://doi.org/10.3727/152599511X13082349958190

Crompton, S. L., & McKay, J. L. (1997). Motives of visitors attending festivals events. *Annals of Tourism Research, 24*(2), 425–439. https://doi.org/10.1016/S0160-7383(97)80010-2

Edensor, T. (2014). The potentialities of light festivals. *Insights, 7*(3). Institute of Advanced Studies: Durham University.

Falassi, A. (1987). Festival: Definition and morphology. In A. Falassi. (Ed.), *Time out of time* (pp. 1–10). University of New Mexico Press.

Falk, J. H., Ballantyne, R., Packer, J., & Benckendorff, P. (2012). Travel and learning: A neglected tourism research area. *Annals of Tourism Research, 39*(2), 908–927. https://doi.org/10.1016/j.annals.2011.11.016

Falk, J. H., & Dierking, L. D. (2000). *Learning from museums: Visitor experiences and the making of meaning.* AltaMira Press.

Falk, J. H., & Storksdieck, M. (2005). Using the contextual model of learning to understand visitors learning from a science centre exhibition. *Science Education, 89*(5), 1–35. https://doi.org/10.1002/sce.20078

Flowers, R., & Swan, E. (2012). Introduction: Why food? Why pedagogy? Why adult education? *Australian Journal of Adult Learning, 52*(3), 419–433. https://search.informit.org/doi/10.3316/ielapa.084521713642071

Flowers, R., & Swan, E. (2015). Food pedagogies, histories, definitions and moralities. In R. Flowers & E. Swan (Eds.), *Food pedagogies* (pp. 1–27). Routledge.

Fontefrancesco, M. F., & Zocchi, D. M. (2020). Revising traditional food knowledge through food festivals. The case of the pink asparagus festival in Mezzago, Italy. *Frontiers in Sustainable Food Systems, 4*, 596028. https://doi.org/10.3389/fsufs.2020.596028

Fyrberg, A., & Jüriado, R. (2009). What about interaction? *Journal of Service Management, 20*(4), 420–432. https://doi.org/10.1108/09564230910978511

Getz, D. (2010). The nature and scope of festival studies. *International Journal of Event Management Research, 5*(1), 1–47.

Getz, D. (2012). *Event studies* (2nd ed.). Routledge.

Getz, D., & Robinson, R. N. S. (2014). Foodies and food events. *Scandinavian Journal of Hospitality and Tourism, 14*(3), 315–330. https://doi.org/10.1080/15022250.2014.946227

Grenier, R. S. (2010). All work and no play makes for a dull museum visitor. *New Directions for Adult and Continuing Education, 127*, 77–85. https://doi.org/10.1002/ace.383

Heim, A. B., & Holt, E. A. (2022). Undergraduates' motivation following a zoo experience: Status matters but structure does not. *Journal of Experiential Education, 45*(1), 68–87. https://doi.org/10.1177/10538259211012716

Horng, J.-S., Su, C.-S., & So, S.-I.-A. (2013). Segmenting food festival visitors: Applying the theory of planned behaviour and lifestyle. *Journal of Convention & Event Tourism, 14*(3), 193–216. https://doi.org/10.1080/15470148.2013.814038

Ind, N., Iglesias, O., & Schultz, M. (2013). Building brands together: Emergence and outcomes of co-creation. *California Management Review, 55*(3), 5–26. https://doi.org/10.1525/cmr.2013.55.3.5

Jaimangal-Jones, D. (2014). Utilising ethnography and participant observations in festival and event research. *International Journal of Event and Festival Management*, *5*(1), 39–55. https://doi.org/10.1108/IJEFM-09-2012-0030

Kajzer Mitchell, I. Low, W., Davenport, E., & Brigham, T. (2017). Running wild in the marketplace: The articulation and negotiation of an alternative food network. *Journal of Marketing Management*, *33*(7–8), 502–528. https://doi.org/10.1080/0267257X.2017.1329224

Kanhadilok, P., & Watts, M. (2014). Adult play-learning: Observing informal family education at a science museum. *Studies in the Education of Adults*, *46*(1), 23–41. https://doi.org/10.1080/02660830.2014.11661655

Kearns, R. A. (2000). Seeing with clarity: Undertaking observational research. In I. Hay (Ed.), *Qualitative research methods in human geography* (pp. 241–258). Oxford University Press.

Kolb, A. Y., & Kolb, D. A. (2010). Learning to play, playing to learn: A case study of a *ludic* learning space. *Journal of Organizational Change Management*, *23*(1), 26–50. https://doi.org/10.1108/09534811011017199

Kolb, D. A. (1984). *Experiential learning: Experience as the source of learning and development*. Prentice-Hall.

Lau, C., & Li, Y. (2019). Analyzing the effects of an urban food festival: A place theory approach. *Annals of Tourism Research*, *74*, 43–55. https://doi.org/10.1016/j.annals.2018.10.004

Lefrid, M., & Torres, E. N. (2022). Hungry for food and community: A study of visitors to food and wine festivals. *Journal of Vacation Marketing*, *28*(3), 366–384. https://doi.org/10.1177/13567667211060568

Levkoe, C. Z. (2006). Learning democracy through food justice movements. *Agriculture and Human Values*, *23*(1), 89–98. https://doi.org/10.1007/s10460-005-5871-5

Lusch, R. L., & Nambisan, S. (2015). Service innovation: A service-dominant logic perspective. *MIS Quarterly*, *39*(1), 155–175.

Lusch, R. L., & Vargo, S. L. (2014). *Service-dominant logic: Premises, prospects, possibilities*. Cambridge University Press.

Mackellar, J. (2013). Participant observation at events: Theory, practice and potential. *International Journal of Event and Festival Management*, *4*(1), 56–65. https://doi.org/10.1108/17582951311307511

Merriam Webster (n.d.). *Ludic*. https://www.merriam-webster.com/dictionary/ludic

Neuhofer, B., Celuch, K., & Lind To, T. (2020). Experience design and the dimensions of transformative festival experiences. *International Journal of Contemporary Hospitality Management*, *32*(9), 2881–2901. https://doi.org/10.1108/IJCHM-01-2020-0008

Organ, K., Koenig-Lewis, N., Palmer, A., & Probert, J. (2015). Festivals as agents of behaviour change: A study of food festivals engagement and subsequent food choices. *Tourism Management*, *48*, 84–99. https://doi.org/10.1016/j.tourman.2014.10.021

Packer, J., & Ballantyne, R. (2005). Solitary vs. shared: Exploring the social dimension of museum learning. *Curator: The Museum Journal*, *48*(2), 177–192. https://doi.org/10.1111/j.2151-6952.2005.tb00165.x

Paris, S. G. (1997). Situated motivation and informal learning. *Journal of Museum Education*, *22*(2–3), 22–27. https://doi.org/10.1080/10598650.1997.11510356

Pinho, N., Beirao, G., Patricio, L., & Fisk, R. P. (2014). Understanding value co-creation in complex services with many actors. *Journal of Service Management*, *25*(4), 470–493. https://doi.org/10.1108/JOSM-02-2014-0055

Prahalad, C. K., & Ramaswamy, V. (2004). Co-creation experiences: The next practice in value creation. *Journal of Interactive Marketing*, *18*(3), 5–14. https://doi.org/10.1002/dir.20015

Rachão, S. A. S., Breda, Z., Fernandes, C., & Joukes, V. (2020). Food and wine experiences towards co-creation in tourism. *Tourism Review*, *76*(5), 1050–1066. https://doi.org/10.1108/TR-01-2019-0026

Rachão, S. A. S., Breda, Z., Fernandes, C., & Joukes, V. (2021). Drivers of experience co-creation in food and wine tourism: An exploratory quantitative analysis. *Tourism Management Perspectives*, *37*, 100783. https://doi.org/10.1016/j.tmp.2020.100783

Richards, G. (2015). Evolving gastronomic experiences: From food to foodies to foodscapes. *Journal of Gastronomy and Tourism*, *1*(1), 1–4. https://doi.org/10.3727/216929715X14298190828796

Richards, G. (2021). Evolving research perspectives on food and gastronomic experiences in tourism. *International Journal of Contemporary Hospitality Management*, *33*(3), 1037–1058. https://doi.org/10.1108/IJCHM-10-2020-1217

Richards, G., & Raymon, C. (2000). Creative tourism. *ATLAS News*, *23*, 16–20.

Rossetti, G., & Quinn, B. (2019). Learning at literary festivals. In I. Jenkins & K. A. Lund (Eds.), *Literary tourism: Theories, practice and case studies* (pp. 93–105). CABI.

Smith, P. K., & Pellegrini, A. (2008). Learning through play. In *Encyclopedia on Early Childhood Development*. CEECD/SKC-ECD. https://www.child-encyclopedia.com/play/according-experts/learning-through-play

Spence, C. (2022). Interacting with food: Tasting with the hands. *International Journal of Gastronomy and Food Science*, *30*, 100620. https://doi.org/10.1016/j.ijgfs.2022.100620

Sumner, J. (2008). Eating as a pedagogical act: Food as catalyst for adult education for sustainability. *Kursiv – Journal fur politische Bildung*, *4*(1), 23–37.

Tatter, G. (2019). Playing to Learn: How a pedagogy of play can enliven the classroom, for students of all ages. *Research Stories*, Harvard Graduate School of Education. https://www.gse.harvard.edu/news/uk/19/03/playing-learn

Tölkes, C., & Butzmann, E. (2018). Motivating pro-sustainable behaviour: The potential of green events – A case study from the Munich Street Festival. *Sustainability*, *10*(10), 3731. https://doi.org/10.3390/su10103731

Van Winkle, C., & Bueddefeld, J. (2016). Service-dominant logic and the festival experience. *International Journal of Event and Festival Management*, *7*(3), 237–254. https://doi.org/10.1108/IJEFM-12-2015-0046

Van Winkle, C. M., & Lagay, K. (2012). Learning through tourism: The experience of learning from the tourist's perspective. *Studies in Continuing Education*, *34*(3), 339–355. https://doi.org/10.1080/0158037X.2011.646981

Vargo, S. L., & Lusch, R. F. (2011). It's all B2B...and beyond: Toward a systems perspective of the market. *Industrial Marketing Management*, *40*(2), 181–187. https://doi.org/10.1016/j.indmarman.2010.06.026

Vargo, S. L., & Lusch, R. F. (2016). Institutions and axioms: An extension and update of service-dominant logic. *Journal of the Academy of Marketing Science*, *44*(1), 5–23. https://doi.org/10.1007/s11747-015-0456-3

"What is Harvest Moon?" (n.d.). Harvest Moon Festival. https://www.harvestmoonfestival.ca/aboutus

Wieland, H., Polese, L., Vargo, S., & Lusch, R. (2012). Toward a service (eco)systems perspective on value creation. *International Journal of Service Science, Management, Engineering, and Technology*, *3*(3), 12–25. https://doi.org/10.4018/jssmet.2012070102

Wilson, J., Arshed, N., Shaw, E., & Pret, T. (2016). Expanding the domain of festival research: A review and research agenda. *International Journal of Management Reviews, 19*(2), 129–257. https://doi.org/10.1111/ijmr.12093

Wyatt, B., Leask, A., & Barron, P. (2021). Designing dark tourism experiences: An exploration of edutainment interpretation at lighter dark visitor attractions. *Journal of Heritage Tourism, 16*(4), 433–449. https://doi.org/10.1080/1743873X.2020.1858087

14 Learning (with) Negative Emotions

The Future of Science Festivals

Adalberto Fernandes

Introduction

Edutainment is the practice of educating through entertainment, associated also with the introduction of computers and new media technologies in educational settings (Aksakal, 2015; Anikina & Yakimenko, 2015). One well-known example of edutainment practice is the Disney media corporation, a business historically focused on commercialising fictional content for children (Telotte, 2004) and war propaganda (Colorado, 2019; Cras, 2019; Dennis, 2004; Manninen, 2019; Shale, 1976; Watts, 1995), and whose media expertise has been used to convey scientific topics entertainingly and persuasively. Regarding this point, Disney (1945, p. 119) commented,

> The motion picture took a leading part in all phases of wartime education, propaganda and information as well as training. It explained and supported ideas, it showed with impartial fidelity the course of events (…). So successful was the motion picture in this task of education for war that close attention was once more given to its capacity as a means for enlightenment and teaching.

Edutainment is also a proposal that emerges in tandem with the critique of the rigidity of the school system that privileges order and strict norms to achieve an equal learning process for different individuals (Bourdieu et al., 1990; Foucault, 2020; Freire, 2018; Illich, 2022). This involves standardising methods of education, limiting the space for emotions of fun, joy, and laughter – all of which are apparently necessary for successful educative processes according to edutainment scholars as it will be shown later.

Science festivals share common features with edutainment, characterised by promoting public encounters with science in more informal modes. Science festivals can be considered a practice closely associated with edutainment, not just because they appear as an alternative informal practice to the school system but also because of the relevant role that entertainment and emotion play in those festivals (Terzian, 2013). It is possible to say that science festivals have been, historically, practising a kind of "edutainment" *avant la lettre*

DOI: 10.4324/9781003305415-18

(even if less concerned with educating individuals than obtaining positive public attitudes towards science) by using techniques of spectacularisation of scientific and technological achievements (Forgan, 1998).

It is certainly possible to envisage a fruitful research agenda between edutainment and science festivals, given the former can learn with the long history of use of emotion and spectacularisation techniques in making science appealing to the public, while the latter can learn with edutainment how to make learning possible while using entertainment techniques. The relationship between edutainment and science festivals has been studied by some scholars (Chen, 2014), but the division of conceptual work between education sciences (Mohd Yusof et al., 2014), information and communication technologies (Feugère et al., 2013), science communication (Niemann et al., 2020), and leisure and tourism studies (Pratt & Suntikul, 2016) have separated what could otherwise be a more fruitful encounter between edutainment and science festivals literature.

One important dimension that unites edutainment and science festivals, and which needs to be critically accounted for, is their shared position concerning the role of emotion in education and the creation of positive attitudes towards science. They have a common understanding about what kind of emotion is more useful to make science appealing and understandable, and that type of emotion seems to be rarely contested. Edutainment and science festivals rely mostly on the persuasive force of positive emotions, given the appeal they seem to have in the learning process. The literature claims that "edutainment is defined as an application compounded with educational aims and measurements (…) and having a good time" (Aksakal, 2015). Also, "the cognitive process is not required to take place in the formal (and often boring) environment and is able to turn into a wholesome entertainment with the acquisition of knowledge" (Anikina & Yakimenko, 2015). Edutainment is following, thus, what many studies have been consistently highlighting, namely, that an individual is more predisposed to learn if the experience is enjoyable (Strouse et al., 2019).

However, it is necessary to ask if an enlarged pallet of emotions could not also be useful for edutainment and science festival purposes? It is no surprise that subjects are moved by multiple emotions, from pleasant to unpleasant (Lin et al., 2018). In the context of education, it is also known that the process of learning is full of multiple emotions, from the enjoyment of learning something new to the complicated process of trial and error with its associated frustration (Alfi et al., 2004). This means the negative emotions associated with the integral process of subjects' education in the context of edutainment and science festivals should not be hidden. Otherwise, edutainment science festivals can create audiences which are averse to failure. The undesirable outcome of this aversion to the difficulties proper to the educational experience can jeopardise the educational objectives of edutainment and science festivals because these can, inadvertently, hide the existence of failure moments associated with the learning process.

Science festivals and the prevalence of positive emotions

According to the literature, edutainment seems to have a clear position about what constitutes the type of emotion conducive to successful learning, which is focused mostly on positive-related emotions, such as feeling friendly, comfortable, and amazing, so that no one feels restricted or afraid, and can thus interact freely and happily (Feiyue, 2022). As such, edutainment promotes feelings of fun (Ferraz-Caetano & Dias, 2021), excitement (Robertson & Lesser, 2013), enthusiasm (Sorathia & Servidio, 2012), and enjoyment (Bertacchini et al., 2012). Likewise, the literature about science festivals also points towards a prevalence of positive emotions. Science festivals are spaces of celebration of science, they occur in informal places (e.g. shopping malls, parks), and they have interactive and artistic activities, alongside talks and debates tailored for non-specialised publics, focusing especially on families (Bultitude et al., 2011; Jensen & Buckley, 2014; Pappa & Koliopoulos, 2021). In the science festival, a pleasurable environment motivates education through an emotional work of making science appealing, interesting, and enjoyable (Davies, 2019b). Science festivals are organised around the assumption that pleasure and enjoyment are the most relevant drives to knowledge transmission (Canovan, 2020). Also, the literature that focuses simultaneously on edutainment and science festivals tends to privilege the role of positive feelings to achieve educational outcomes. This appears to be done through a strategy of offering a spectacle and an image of a softer science, rather than hard scientific theory (Chen, 2014), in addition to activities that are enjoyable, exciting, and fun (Mohd Yusof et al., 2014). The learning process therefore becomes easier and more entertaining (Feugère et al., 2013), promoting positive experiences and fascination (Pratt & Suntikul, 2016).

The power of negative emotions

Edutainment practices and science festivals seem to be based on a restricted notion of positive emotions, leaving aside negative ones. This means edutainment and science festivals do not cover the entire subjects' emotional palette, which is also composed of negative emotions, such as fear, anger, anxiety, depression, and grief. This is not unusual. The original contribution of edutainment in science festivals is to make science and education more pleasurable, which seems to be best achieved through positive emotions. This way of approaching emotion may, however, create an inflated expectation among learners (and science festival attendees) that the process of learning should always be colourful and fun, and that they can acquire information without work and serious study (Okan, 2003). It can be also claimed that promoting mostly positive feelings through edutainment in science festivals may downplay the importance negative feelings may have in the learning process.

In review of the above, four arguments may be offered to expand the concept of emotion used by edutainment in science festivals to include negative feelings that may contribute to the learning process. Firstly, the use of negative feelings is a way of being more faithful to the process of education, which can be marked by the challenges associated with the trial-and-error experience. Secondly, an exclusive reliance on positive feelings can elude the learner of the real difficulties of learning and make her more frustrated. Thirdly, to discuss negative emotions during science festivals may appeal to attendees that have negative feelings associated with the learning process. Finally, negative emotions can be a way to critically approach the moments in which science becomes dangerous and, therefore, associated with negative feelings.

In review of the first argument, negative feelings are intrinsic to scientific production (Wright et al., 2011) and the process of science learning (King et al., 2017), both marked by the possibility of failure which leads to frustration, anxiety, etc. To produce and to learn science, through experience, is to be prone to failure because they are not activities always programmed to invariably work out. This failure is, nonetheless, instructive, as it teaches that some paths are wrong and, in comparison, may indicate where the correct paths are. This means failure and its associated negative feelings should be taken as normal in the learning process (Alfi et al., 2004).

Regarding the second argument, an edutainment science festival that only promises happiness has fewer resources to handle the frustration that can inevitably arise in the learning experience, given the learners' idiosyncrasies, who do not respond the same way to edutainment science festivals (Niemann et al., 2020). The over-reliance on positive emotions for education may create a false expectative in the learner (Okan, 2003), which can augment its disaffection towards learning, instead of preparing her for negative feelings linked to failure, which are likely to happen, given that learning is an open experience, not a program destined to always run successfully.

Learners with negative past experiences with education know what makes the learning process difficult and uninteresting. They have a rich emotional knowledge, and thus with reference to the third argument, these attendees should have a say in the construction of edutainment science festivals to attract those individuals that relate to those negative feelings. This is nothing more than the adaptation of the science communication strategy of using in-group subjects to convince similar others to join opposite groups (Fielding et al., 2020). In analogous terms, those excluded from the learning experience, because they have negative feelings towards it, have important insights that could help in the organisation of edutainment science festivals to appeal to those that are also experiencing negative learning experiences. One of the persistent challenges of science festivals is the fact that they tend to appeal to subjects that are already interested in science or have a scientific background (Kennedy et al., 2018), albeit their investment in spectacular techniques and positive emotions to deliver science to diverse publics (Davies, 2019b). This means that the expansion of emotions that can be experienced, including negative ones, may create new opportunities for a relationship with science (Humm et al., 2020).

Finally, if people are taught that knowledge is associated with powerful techno-scientific institutions, market interests, risks, and controversies (Rose et al., 2017), they can use the negative feelings those associations provoke to be more critical of the dangers associated with techno-scientific development. In this way, and in support of the fourth argument, science education and the creation of positive public attitudes towards science ceases to be an exclusive "end in itself" of science festivals (Davies, 2019a). Using a mixture of knowledge and mixed feelings is therefore important when using edutainment techniques in science festivals.

Horror Science-Fiction Movie Festival: A theoretical proposal

It can be said that negative emotions are contrary to the use of edutainment in science festival agendas, as the traditional purpose is to use entertainment and positive emotions to make the science festival learning process more appealing. However, negative emotions can be entertaining too. This final argumentative move is necessary because to say that negative feelings are part of the learning process is not the same as claiming that they are entertaining. In other words, it is not sufficient to claim that negative feelings are constitutive of the learning process. It is also necessary to defend that they can be as entertaining as positive feelings. Only in this way can this theoretical chapter contribute to edutainment within science festivals.

The association between negative feelings and science, in terms of making it more appealing, is not new. Some exhibitions at science centres and museums (e.g. Science Gallery, Dublin), the *Horrible Science* best-selling book series (Bell, 2011), and Halloween-themed science communication projects are examples that use the power of negative feelings to educate or to make science more attractive. While not related to science, "dark tourism" studies have already shown the appeal of suffering and tragedy sites as places of edutainment (Wyatt et al., 2021). However, none of these examples refers to science festivals; they also do not explicitly have edutainment objectives of making people acquire knowledge, nor do they address the frustration of the learning process as a topic that can connect with disaffected learners. There is, therefore, no edutainment science festival based on entertainment with negative feelings about education for learning purposes in the current literature. An example of such science festivals are Horror Science-Fiction Movie Festivals, which may offer an edutainment science festival experience based on negative emotions.

Horror film festivals, which are of sci-fi popular culture, are not necessarily intended as edutainment science festivals. However, the pseudo or quasi-scientific themes of sci-fi, combined with the scary nature of horror narratives, are appealing to their audiences. There is, thus, an opportunity for edutainment-based science festivals to explore the benefits of using negative emotions. One source of inspiration can be the International Horror and Sci-Fi Film Festival (2022), which, held since 2005, joins horror, science, and fiction with an explicit aim to elicit emotions of fear, anxiety, spookiness, and fright – emotions that are captivating but not necessarily based on positive emotions (Lin & Xu, 2017; Tudor, 1997). Horror film festivals also create a community around a set of negative feelings that are constantly left out of the mainstream entertainment industry (Hills, 2009). Drawing on horror sci-fi festivals, which attract people by appealing to their feelings of fear and anxiety, edutainment-based science festivals of the future can capitalise on a wider pallet of educational emotions to include negative ones, which, as demonstrated, can be part of the entertainment and learning process.

Future work should be in the hands of scholars and practitioners of edutainment science festivals to think and use the negative feelings appropriate to the learning process and make them entertaining for learners. This is the task that lies ahead: an edutainment science festival capable of using the paradox of horror – when fear becomes a positive emotion (Bantinaki, 2012). Given the scarcity of such an event, which does not exist yet because the literature and practice of edutainment science festivals have been investing most of their efforts in positive emotions, this chapter is an attempt to start a dialogue about the value of negative feelings for entertainment and educational purposes. The healthiness of a field of studies is measured not only by its empirical studies but also by its theoretical auto-reflexive capacity to envisage what does not exist yet to guide future empirical work. This theoretical proposal is therefore, at the same time, an ambitious and speculative push into that future direction, revealing also how edutainment and science festivals scholarship is a process where failure may be also instructive.

Conclusion

According to the literature, edutainment and science festivals tend to use a reduced emotional palette to convey science, leaving aside the potential of negative emotions. The experiences of failure, frustration, anxiety, and despair, which are part of learning processes, should be valued to open the space for individual and collective modes of existence prepared to handle those experiences. Otherwise, edutainment science festivals, which are solely focused on positive emotions, can create false and unrealistic expectations in the learner. It is proposed, theoretically, an edutainment science festival to be built in the future around the power of negative emotions, inspired by horror sci-fi movie festivals' appeal, and close to the complicated experience of learning.

References

Aksakal, N. (2015). Theoretical view to the approach of the edutainment. *Procedia – Social and Behavioral Sciences, 186*, 1232–1239. https://doi.org/10.1016/j.sbspro.2015.04.081
Alfi, O., Assor, A., & Katz, I. (2004). Learning to allow temporary failure: Potential benefits, supportive practices and teacher concerns. *Journal of Education for Teaching, 30*(1), 27–41. https://doi.org/10.1080/0260747032000162299

Anikina, O. V., & Yakimenko, E. V. (2015). Edutainment as a modern technology of education. *Procedia – Social and Behavioral Sciences*, *166*, 475–479. https://doi.org/10.1016/j.sbspro.2014.12.558

Bantinaki, K. (2012). The paradox of horror: Fear as a positive emotion. *The Journal of Aesthetics and Art Criticism*, *70*(4), 383–392. https://doi.org/10.1111/j.1540-6245.2012.01530.x

Bell, A. R. (2011). Science as 'Horrible': Irreverent deference in science communication. *Science as Culture*, *20*(4), 491–512. https://doi.org/10.1080/09505431.2011.605921

Bertacchini, F., Bilotta, E., Pantano, P., & Tavernise, A. (2012). Motivating the learning of science topics in secondary school: A constructivist edutainment setting for studying Chaos. *Computers & Education*, *59*(4), 1377–1386. https://doi.org/10.1016/j.compedu.2012.05.001

Bourdieu, P., & Passeron, J.-C. (1990). *Reproduction in Education, Society and Culture*. SAGE Publications.

Bultitude, K., McDonald, D., & Custead, S. (2011). The rise and rise of science festivals: An international review of organised events to celebrate science. *International Journal of Science Education, Part B*, *1*(2), 165–188.

Canovan, C. (2020). More than a grand day out? Learning on school trips to science festivals from the perspectives of teachers, pupils and organisers. *International Journal of Science Education, Part B*, *10*(1), 1–16.

Chen, G. (2014). National Science Festival of Thailand: Historical roots, current activities and future plans of the National Science Fair. *Journal of Science Communication*, *13*(4), C04. https://doi.org/10.22323/2.13040304

Colorado, J. (2019). Racism and popular culture in World War II, from Dr. Seuss to Walt Disney. *OSR Journal of Student Research*, *5*(1). https://scholarworks.lib.csusb.edu/osr/vol5/iss1/165

Cras, P. (2019). Between "good neighbor policy and propaganda animated movies": Walt Disney goes to war (1941–1948). *Revue francaise detudes americaines*, *158*(1), 111–131.

Davies, S. R. (2019a). Science communication is not an end in itself: (Dis)assembling the science festival. *International Journal of Science Education Part B-Communication and Public Engagement*, *9*(1), 1. https://doi.org/10.1080/21548455.2018.1540898

Davies, S. R. (2019b). Science communication as emotion work: Negotiating curiosity and wonder at a science festival. *Science as Culture*, *28*(4), 538–561. https://doi.org/10.1080/09505431.2019.1597035

Dennis, J. P. (2004). The light in the forest is love: Cold war masculinity and the Disney adventure boys. *Americana: The Journal of American Popular Culture, 1900 to Present*, *3*(1). https://www.proquest.com/docview/1519963171/abstract/AFAFCD7759BA48A1PQ/1

Disney, W. (1945). Mickey as professor. *Public Opinion Quarterly*, *9*(2), 119–125. https://doi.org/10.1086/265726

Feiyue, Z. (2022). Edutainment methods in the learning process: Quickly, fun and satisfying. *International Journal of Environment, Engineering and Education*, *4*(1), 1. https://doi.org/10.55151/ijeedu.v4i1.41

Ferraz-Caetano, J., & Dias, D. (2021). Chemistry edutainment: A storytelling activity for middle-school children. In L. C. Mihăeş, R. Andreescu, & A. Dimitriu (Eds.), *Advances in linguistics and communication studies* (pp. 364–391). IGI Global.

Feugère, L., d'Alessandro, C., & Doval, B. (2013). Performative voice synthesis for edutainment in acoustic phonetics and singing: A case study using the "Cantor

Digitalis". In M. Mancas, N. d'Alessandro, X. Siebert, B. Gosselin, C. Valderrama, & T. Dutoit (Eds.), *Intelligent technologies for interactive entertainment* (pp. 169–178). Springer International Publishing.

Fielding, K. S., Hornsey, M. J., Thai, H. A., & Toh, L. L. (2020). Using ingroup messengers and ingroup values to promote climate change policy. *Climatic Change*, *158*(2), 181–199. https://doi.org/10.1007/s10584-019-02561-z

Forgan, S. (1998). Festivals of science and the two cultures: Science, design and display in the Festival of Britain, 1951. *The British Journal for the History of Science, 31*(2), 217–240. https://doi.org/10.1017/S0007087498003264

Foucault, F. (2020). *Discipline and punish: The birth of the prison*. Penguin.

Freire, P. (2018). *Pedagogy of the oppressed* (4th ed.). Bloomsbury Academic.

Hills, M. J. (2009). Attending horror film festivals and conventions: Liveness, subcultural capital and 'flesh-and-blood genre communities'. In I. Conrich (Ed.), *Horror zone: Entering the world of contemporary horror cinema: The cultural experience of contemporary horror cinema* (pp. 87–102). I. B. Tauris.

Humm, C., Schrögel, P., & Leßmöllmann, A. (2020). Feeling left out: Underserved audiences in science communication. *Media and Communication, 8*(1), 1. https://doi.org/10.17645/mac.v8i1.2480

Illich, I. (2022). *Deschooling society*. KKIEN Publ. Int.

International Horror and Sci-Fi Film Festival. (2022). https://en.wikipedia.org/w/index.php?title=International_Horror_and_Sci-Fi_Film_Festival&oldid=1106812885

Jensen, E., & Buckley, N. (2014). Why people attend science festivals: Interests, motivations and self-reported benefits of public engagement with research. *Public Understanding of Science, 23*(5), 557–573.

Kennedy, E. B., Jensen, E. A., & Verbeke, M. (2018). Preaching to the scientifically converted: Evaluating inclusivity in science festival audiences. *International Journal of Science Education, Part B, 8*(1), 14–21. https://doi.org/10.1080/21548455.2017.1371356

King, D., Ritchie, S. M., Sandhu, M., Henderson, S., & Boland, B. (2017). Temporality of emotion: Antecedent and successive variants of frustration when learning chemistry. *Science Education, 101*(4), 639–672. https://doi.org/10.1002/sce.21277

Lin, C., & Xu, Z. (2017). Watching TV series with horror content: Audience attributes, motivations, involvement and enjoyment. *Journal of Broadcasting & Electronic Media, 61*(4), 638–657. https://doi.org/10.1080/08838151.2017.1375503

Lin, J.-H. T., Wu, D.-Y., & Tao, C.-C. (2018). So scary, yet so fun: The role of self-efficacy in enjoyment of a virtual reality horror game. *New Media & Society, 20*(9), 3223–3242. https://doi.org/10.1177/1461444817744850

Manninen, T. (2019). "Let Slip the (Donald) Ducks of War!"—Ethical considerations about Disney's war propaganda. In R. Davis (Ed.), *Disney and philosophy* (pp. 217–226). John Wiley & Sons, Ltd.

Mohd Yusof, A., Daniel, E. G. S., Low, W. Y., & Ab. Aziz, K. (2014). Teachers' perception of mobile edutainment for special needs learners: The Malaysian case. *International Journal of Inclusive Education, 18*(12), 1237–1246. https://doi.org/10.1080/13603116.2014.885595

Niemann, P., Bittner, L., Schrögel, P., & Hauser, C. (2020). Science slams as edutainment: A reception study. *Media and Communication, 8*(1), 177–190. https://doi.org/10.17645/mac.v8i1.2459

Okan, Z. (2003). Edutainment: Is learning at risk? *British Journal of Educational Technology, 34*(3), 255–264. https://doi.org/10.1111/1467-8535.00325

Pappa, E., & Koliopoulos, D. (2021). Attempts to categorize and evaluate science festivals, a 30-year-old science communication event: The case of Greece. In B. Schiele, X. Liu, & M. W. Bauer (Eds.), *Science cultures in a diverse world: Knowing, sharing, caring* (pp. 77–89). Springer.

Pratt, S., & Suntikul, W. (2016). Can marine wildlife tourism provide an "Edutaining" experience? *Journal of Travel & Tourism Marketing, 33*(6), 867–884. https://doi.org/10.1080/10548408.2015.1069778

Robertson, W., & Lesser, L. M. (2013). Scientific skateboarding and mathematical music: Edutainment that actively engages middle school students. *European Journal of Science and Mathematics Education, 1*(2), 60–68.

Rose, K. M., Korzekwa, K., Brossard, D., Scheufele, D. A., & Heisler, L. (2017). Engaging the public at a science festival: Findings from a panel on human gene editing. *Science Communication, 39*(2), 250–277. https://doi.org/10.1177/1075547017697981

Shale, R. A. (1976). *Donald Duck joins up: The Walt Disney Studio during World War II*. University of Michigan.

Sorathia, K., & Servidio, R. (2012). Learning and experience: Teaching tangible interaction & edutainment. *Procedia – Social and Behavioral Sciences, 64*, 265–274. https://doi.org/10.1016/j.sbspro.2012.11.031

Strouse, G. A., Newland, L. A., & Mourlam, D. J. (2019). Educational and fun? Parent versus preschooler perceptions and co-use of digital and print media. *AERA Open, 5*(3). https://doi.org/10.1177/2332858419861085

Telotte, J. P. (2004). Minor hazards: Disney and the color adventure. *Quarterly Review of Film and Video, 21*(4), 273–281. https://doi.org/10.1080/10509200490446150

Terzian, S. (2013). *Science education and citizenship: Fairs, clubs, and talent searches for American youth, 1918–1958*. Palgrave Macmillan.

Tudor, A. (1997). Why horror? The peculiar pleasures of a popular genre. *Cultural Studies, 11*(3), 443–463. https://doi.org/10.1080/095023897335691

Watts, S. (1995). Walt Disney: Art and politics in the American century. *Journal of American History, 82*(1), 84–110. https://doi.org/10.2307/2081916

Wright, J. G., Khetani, N., & Stephens, D. (2011). Burnout among faculty physicians in an academic health science centre. *Paediatrics & Child Health, 16*(7), 409–413. https://doi.org/10.1093/pch/16.7.409

Wyatt, B., Leask, A., & Barron, P. (2021). Designing dark tourism experiences: An exploration of edutainment interpretation at lighter dark visitor attractions. *Journal of Heritage Tourism, 16*(4), 433–449. https://doi.org/10.1080/1743873X.2020.1858087

15 Afterword

The Future of Edutainment in Festival Experiences

Brianna Wyatt and Giulia Rossetti

Ideation

The ideation underpinning this book stems from the editors' previous works on festivals and edutainment experiences, in which gaps in both knowledge and published research were revealed, prompting a need for greater attention on how these two topics work together to create enhanced audience experiences. Drawing on her areas of expertise – developed from her PhD research (Rossetti, 2020) that explored the learning and cultural impacts of participating in literary festivals in Italy and Ireland – Giulia formed the initial ideas for the book, which had originally focused on understanding how people learn, acquire culture, and communicate at festivals. This idea was further inspired by her subsequent research that focused on learning at literary festivals (Rossetti & Quinn, 2019), how cultural capital development occurs at rural festivals (Rossetti & Quinn, 2021, 2022) and how events can be transformational and can change lives (Antchak et al., 2022) – all which came to a deliver a revelation that there existed an under-developed knowledge of how education and entertainment work together to create enhanced festival experiences. In discussing these findings with Brianna, whose areas of expertise are edutainment interpretation and experience design – developed from her PhD research (Wyatt, 2019) and subsequent publications on edutainment experiences (Wyatt et al., 2021), interpretation design (Wyatt, 2022), and re-enactment experiences (Wyatt et al., 2023) – it was realised that many festivals do employ a variety of edutainment methods. But in discussing the existing publications (or lack thereof) on edutaining festival experiences, it was soon revealed that more needed to be known. Taking these revelations to Jane, who has explored a range of festival experiences, including festival impacts beyond the financial ones (Carlsen et al., 2007), creating fringe atmospheres (Frew & Ali-Knight, 2009), place-making at festivals (Platt & Ali-Knight, 2018) and using the senses in co-producing dementia events (Stewart et al., 2022), it was soon determined that a curated collection of case studies on edutainment within festivals was needed and would be significant for enhancing understanding across academia and practice. The results of this determination have proven successful since the contributing chapters have highlighted varying ways in

DOI: 10.4324/9781003305415-19

which edutainment is applied across a wide range of festival experiences, thus prompting further discussion for how it can be applied further in the future.

Reflecting on edutainment in festival experiences

This book has shown that edutainment allows for enhanced audience experiences that deeply engage festival audiences. As the contributors to Part I of this book – *Planning edutainment* – have demonstrated, strategically planned and designed edutainment experiences can prompt festival attendees to become inspired to take their learning from the festival experience and apply it at home in an extended learning experience. Involving local businesses and attractions can help to extend the festival experience and therefore learning opportunities, while theme parks and attractions can host their own festival experiences to extend their offerings and enhance their visitors' experience. Moreover, involving the local community in festival experiences and applying edutainment through hands-on activities can help festival attendees to recognise what they are capable of, enhance their self-belief, and grow their understanding through learning as play. Such experiences are thus capable of creating extremely memorable experiences. However, as the contributors of Part II – *Audience engagement* – demonstrated, these experiences are not confined to physical settings. Thanks to advances in technology, festivals now have the ability to create cross-border experiences and widen their audience participation through online platforms. In many cases, these types of experiences may be the only experience attendees are capable of having, and thus providing such experiences digitally is not only a benefit for accessibility but also inclusivity. Beyond the digital realm, offering edutainment experiences for children and adults alike, as well as kid-focused activities, and after-hours adult-focused activities are proven ways festivals can further engage their audiences across multiple levels and thus stimulate deeper learning and enjoyment. Moreover, community involvement within the festival experience through strategically designed edutainment activities can help to both educate and entertain audiences but also strengthen the social bonds between attendees and the local community hosting the festival experience.

Prominent themes in all leisure-related studies at the moment are sustainability and EDI (equality, diversity, and inclusivity), which is a reflection of society's focus on creating a better tomorrow. In Part III of this book – *Sustainability and EDI* – the contributors make clear that the more interactive and co-creative the festival experience is, through, for example, edutaining experiential learning, the more engaged attendees will be and the more they will learn and grow to appreciate the festival's message – such as sustainability. Importantly, it is demonstrated that to truly engage audiences with such important topics, the mission of being sustainable and/or inclusive, diverse, and accessible must be at the forefront of the festival agenda and experience design. Doing this is shown to be quite easy if employing technology, for example, media and film that can enhance the edutainment activities being offered.

Such designs must therefore consider the theme of the festival experience and story being told, which in Part IV – *Experiencing edutainment* – the contributors highlight as being of the utmost importance. Specifically, drawing on the emotional connections attendees may have with the festival theme or storyline through edutainment activities, such as re-enactment, can help to create more memorable experiences. Through experiential learning experiences designed around themes and storylines, attendees can learn more deeply. However, offering them time to reflect on their experiences while still in the festival experience is certainly a benefit to the learning experience. What is more, recognising how audiences engage and experience festivals is important to consider when planning and designing audience experiences. However, also considering why some may not engage with particular festivals or festival themes is important, as festival organisers can use this knowledge to redesign their promotional strategies and experience offerings so that those negative emotions or perspectives towards certain festival themes may be potentially turned into positive ones.

The future of edutainment in festival experiences

Given the information and findings the book's contributors have provided, it is clear that the future of edutainment in festival experiences will likely be driven by technology, co-creation and immersion. It is likely that edutainment methods, such as thematic designs, sensory stimulation, co-creative activities, re-enactments, and varying media, such as film and photography, will continue to engage festival audiences as we move into the future. However, it is uncertain if virtual festival experiences through live streaming and online engagement platforms will continue to thrive. As society moves further away from the lockdown experiences of the COVID-19 pandemic, there is a return of a desire to experience festivals first-hand and to be able to say "I was there". These first-hand experiences offer what Garcia (2021) calls "life-defining moments of shared memories". Yet, as we've seen through the breadth of studies covered in this book, many festival experiences can still engage audiences in a virtual way, not to mention it helps to lower geographic boundaries and increase accessibility and inclusivity. In fact, numerous tourism experiences have capitalised on the virtual experience, offering live-streamed and pre-recorded guided tours of their premises in order to reach those visitors who may not have the means to travel and experience them in person.

As we move further into the future, society is expected to maintain its desire for interactive, immersive, and hyper-real experiences. This is likely because society has become largely dependent on technology and media (Perry, 2020), which, in turn, has increased their expectations for more elaborate and simulative experiences (Neuhofer & Buhalis, 2014). Some have suggested AR/VR and wearable technologies, such as headgear, glasses, and gloves, will become a sought-after edutainment method in tourism, as these will help to enhance experiences and allow full immersion (Wright, 2021). Such technologies are likely to soon be seen in festival experiences, both in at-home and on-site experiences, as attendees can experience the festival live from their home through VR headsets, while attendees on-site can wear AR glasses to enhance the scenery as they move around the venue space. Simulation experiences are also a future possibility for festivals, especially for those that have a strong storyline or message for attendees to learn about. Three-hundred-and-sixty-degree panoramic films or simulation rides like Soarin' Around the World at Disney's California Adventure in the United States may offer such simulation experiences (Wu et al., 2020). While the costs of running such experiences may be too high for a one-off or small community festival, larger and annual festivals that have permanent venue spaces, such as Glastonbury or the Edinburgh International Festival, may benefit from such technologies in their attempts to both educate and entertain audiences.

In addition to technology, well-being, social bonding, and building cultural capital will likely be dominant themes within festival experiences in the future, and the use of edutainment will likely be a driving force for delivering those social messages and missions. As many of the book's contributors have demonstrated, festivals are inherently underpinned with cultural value and meaning that can help to build social connections between attendees. In fact, scholars have suggested festival participation and involvement – that being physical, emotional, intellectual, and/or social – is a way for festival attendees to enhance their social relationships (Wilks & Quinn, 2016) and cultural capital (Rossetti & Quinn, 2021). Future festivals can continue to rely on their programming, marketing, and promotions, and general edutainment activities to build such relationships. However, embracing some of the aforementioned technologies and allowing for more immersive and/or simulated experiences that attendees can engage with through active involvement across multiple levels is arguably an effective means of boosting social bonding. As attendees physically engage with, for example, sensory-stimulating activities, together – whether that be in their own groups or with other attendees they may not know – they will certainly be able to enhance their cultural capital. In addition, drawing attendees further into the storylines, narratives, and/or festival message and mission through interactivity and group experiential edutainment activities will help them to become more engrossed with those narratives thus feeling a stronger connection to it and those around them, which may, in turn, create "genuine interest in learning and thinking more deeply about the world" (Rossetti & Quinn, 2021, p. 48).

Research futures for edutainment in festival experiences

Suggestions for future studies have been highlighted throughout the book, and they include recommendations for a better understanding of how the concept of festival edutainment is defined and used by varying stakeholders (see Chapter 5) and in non-traditional festival settings (see Chapter 4). They also include suggestions to explore different methodological approaches to investigate edutainment at festivals and events (see Chapter 11), investigate global comparisons of the types of activities programmed at edutaining festivals (see Chapter 7), and analyse long-term outcomes of edutainment activities at festivals (see Chapter 13). Other suggestions concern the exploration of certain topics in relation to edutainment in festival settings. For instance, there are calls for further investigations of how the past is told in edutaining festival experiences and how authenticity is therefore managed via edutainment at festivals (see Chapters 3 and 12). Other suggestions include exploring how festival edutainment is designed and evaluated in relation to sustainability (see Chapter 9) and community and social bonding (see Chapter 8), while others called for greater understanding of the use of technology and gamification for festival edutainment (see Chapters 6 and 10), and an investigation into the different types of learners for edutaining festival activities (see Chapter 13). Notably, it has been recommended that future studies should better explore the role of negative emotions in planning and consuming edutaining festival experiences (see Chapter 14), which would contribute to the growing literature pertaining to audience motivations and/or other leisure experiences that can induce negative emotions, such as dark tourism. Such recommendations help to enhance those put forward in Chapter 2, which calls for more higher education institutions to work with science festival organisers in order to better support the development of 'science capital'.

True, scholarly interest in the aforementioned future themes has slowly started to emerge, with, for example, some exploring how events and festivals generate subjective and community well-being, health, and quality of life (Ahn, 2021; Jepson & Walters, 2021; Rossetti, 2021). However, much more is needed to fully understand the links between learning while having fun (edutainment) at festivals and the promotion of attendees' psychological well-being. For instance, is the enjoyment of learning perceived as an achievement? If so, why? How is edutainment at festivals meaningful for attendees and in what ways can technology and co-creative activities enhance that meaningful perception? Such topics have been explored within tourism research (see e.g. de Groot, 2016; Light & Ivanova, 2022; Magelssen, 2006; Wyatt et al., 2023), but it remains an under-developed focus for scholarly attention within the festival experiencescape.

At festivals, attendees can develop their knowledge and skills, which can lead to values and behavioural changes (Rossetti & Quinn, 2021). Yet, they can also co-develop these same skills and understanding to create shared values and social change as several of this book's contributors have demonstrated. As such, future research must further investigate cultural capital development both during and after the festival experience. Future studies should also explore the links between building cultural capital at festivals and well-being outcomes, and how edutainment through, for example, advanced technologies can help to boost that development. In addition, they should investigate how learning and edutaining activities impact on audiences' self-esteem, identity, and achievement, which are all components of well-being (Seligman, 2011). Moreover, since edutaining festival experiences are understood as social occasions, where

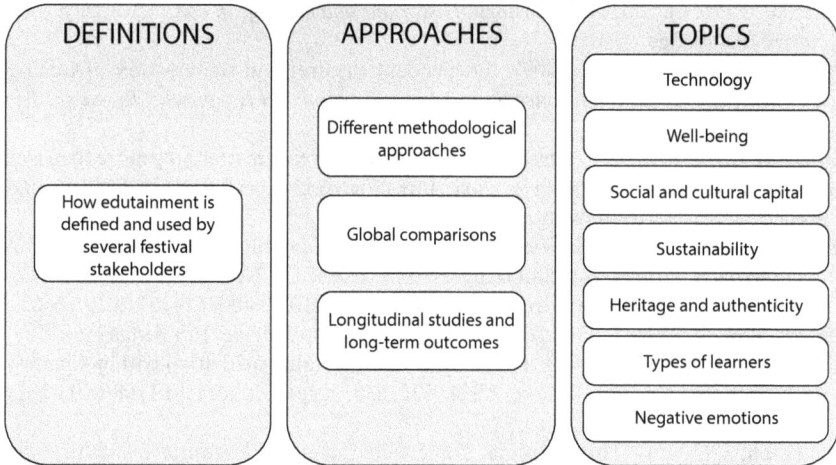

Figure 15.1 Research futures for edutainment in festival experiences.
Source: (Authors).

audiences can strengthen their social networks and increase their social relationships (Hassanli et al., 2020; Wilks & Quinn, 2016), future research should explore how the bridging and bonding of social capital occurs while engaging in edutaining festival experiences.

Given all these revelations, Figure 15.1 summarises the key areas for future research that the editors and contributing authors of this book have identified. Edutainment must be better defined in relation to its use within festival experiences. Research approaches to explore and investigate the use of edutainment within festival experiences also deserve similar scrutiny as that within the tourism sector; this includes the types of methodological approaches taken to explore and investigate this phenomenon. Finally, the range of themes, key terms and topics addressed throughout this book and highlighted in this concluding chapter require a narrowed focus within scholarly outputs. Such studies could then help to extend knowledge and understanding of the benefit for, and use of, edutainment within not just tourism experiences, but importantly, festival experiences.

References

Ahn, Y-J. (2021). Do informal social ties and local festival participation relate to subjective well-being? *International Journal of Environmental Research and Public Health*, *18*(1), 16. https://doi.org/10.3390/ijerph18010016

Antchak, V., Gorchakova, V., & Rossetti, G. (2022). The value of events in times of uncertainty: Insights from balcony performances in Italy during the COVID-19 lockdown. *Annals of Leisure Research*. https://doi.org/10.1080/11745398.2022.2046117

Carlsen, J., Robertson, M., & Ali-Knight, J. (2007). Festivals and events: Beyond economic impacts. *Event Management*, *11*(1/2), 1–98.

de Groot, J. (2016). *Consuming history: Historians and heritage in contemporary popular culture*. Routledge.

Frew, E., & Ali-Knight, J. (2009). Independent theatres and the creation of a fringe atmosphere. *International Journal of Culture, Tourism and Hospitality Research, 3*(3), 211–227.

Garcia, B. (2 August 2021). Why festivals and special events matter now more than ever. *UKRI: UK Research and Innovation*. https://www.ukri.org/blog/why-festivals-and-special-events-matter-now-more-than-ever/

Hassanli, N., Walters, T., & Friedmann, R. (2020). Can cultural festivals function as counterspaces for migrants and refugees? The case of the New Beginnings Festival in Sydney. *Leisure Studies, 39*, 165–180. https://doi.org/10.1080/02614367.2019.1666296

Jepson, A. S., & Walters, T. (Eds.). (2021). *Events and well-being*. Routledge.

Light, D., & Ivanova, P. (2022). Thanatopsis and mortality mediation within "lightest" dark tourism. *Tourism Review, 77*(2), 622–635. https://doi.org/10.1108/tr-03-2021-0106

Magelssen, S. (2006). "This is drama. You are characters": The tourist as fugitive slave in Conner Prairie's "Follow the North Star". *Theatre Topics, 16*(1), 19–34. https://doi.org/10.1353/tt.2006.0011

Neuhofer, B., & Buhalis, D. (2014). Experience, co-creation and technology: Issues, challenges and trends for technology enhanced tourism experiences. In S. McCabe (Ed.), *The Routledge handbook of tourism marketing* (pp. 124–139). Routledge.

Perry, R. E. (2020). The Holocaust is present: Reenacting the Holocaust, then and now. *Holocaust Studies, 26*(2), 152–180. https://doi.org/10.1080/17504902.2019.1578460

Platt, L., & Ali-Knight, J. (2018). Guest editorial. *Journal of Place Management and Development, 11*(3), 262–265.

Rossetti, G. (2020). *Literary festival participation and the development of cultural capital: An analysis of one Irish and one Italian case* [Doctoral Thesis, Technological University Dublin].

Rossetti, G. (2021). The role of literary festival attendance in generating attendees' health and well-being. *International Journal of Event and Festival Management, 12*(3), 265–278. https://doi.org/10.1108/IJEFM-12-2020-0083

Rossetti, G., & Quinn, B. (2019). Insights into the learning dimensions of literary festival experiences. In I. Jenkins & A. K. Lund (Eds.), *Literary tourism: Theories, practice and case studies*. CABI.

Rossetti, G., & Quinn, B. (2021). Understanding the cultural potential of rural festivals: A conceptual framework of cultural capital development. *Journal of Rural Studies, 86*, 46–53. https://doi.org/10.1016/j.jrurstud.2021.05.009

Rossetti, G., & Quinn, B. (2022). The value of the serious leisure perspective in understanding cultural capital embodiment in festival settings. *The Sociological Review*. https://doi.org/10.1177/00380261221108589

Seligman, M. E. (2011). *Flourish: The new positive psychology and the search for well-being*. Free Press.

Stewart, H., Ali-Knight, J., Stephen, S., & Kerr, G. (2022). The 'Senses Framework': A relationship-centred approach to co-producing dementia events in order to allow people to live well after a dementia diagnosis. *Event Management, 26*(1), 157–175. https://doi.org/10.3727/152599521X16192004803683

Wilks, L., & Quinn, B. (2016). Linking social capital, cultural capital and heterotopia at the folk festival. *Journal of Comparative Research in Anthropology and Sociology, 7*, 23–39. https://doi.org/10.21427/D7BP6G

Wright, D. W. M. (2021) Immersive dark tourism experiences: Storytelling at dark tourism attractions in the age of 'the immersive death'. In M. H. Jacobsen (Ed.), *The age of spectacular death* (pp. 89–109). Routledge.

Wu, H. C., Ai, C. H., & Cheng, C. C. (2020). Virtual reality experiences, attachment and experiential outcomes in tourism. *Tourism Review, 75*(3), 481–495. https://doi.org/10.1108/TR-06-2019-0205

Wyatt, B. (2019). *Influences on interpretation: A critical evaluation of the influences on the design and management of interpretation at lighter dark visitor attractions* [Doctoral dissertation, Edinburgh Napier University].

Wyatt, B. (2022). Interpretation design and management: Creating dark edutainment experiences. In D. Agapito, A. Ribeiro, & K. Woosnam (Eds.), *Handbook of tourist experience: Design marketing and management*. Edward Elgar Publishing.

Wyatt, B., Leask, A., & Barron, P. (2021). Designing dark tourism experiences: An exploration of edutainment interpretation at lighter dark visitor attractions. *Journal of Heritage Tourism, 16*(4), 433–449. https://doi.org/10.1080/1743873X.2020.1858087

Wyatt, B., Leask, A., & Barron, P. (2023). Re-enactment in lighter dark tourism: An exploration of re-enactor tour guides and their perspectives on creating visitor experiences. *Journal of Travel Research*. OnlineFirst. https://doi.org/10.1177/00472875221151074

Index

Pages in *italics* refer to figures and pages in **bold** refer to tables.

For Product Safety Concerns and Information please contact our EU
representative GPSR@taylorandfrancis.com
Taylor & Francis Verlag GmbH, Kaufingerstraße 24, 80331 München, Germany

www.ingramcontent.com/pod-product-compliance
Lightning Source LLC
Chambersburg PA
CBHW060256220326
41598CB00027B/4118